The Great Dialogue
of
Nature and Space

The Great Dialogue
of
Nature and Space

Yves Simon

Edited by Gerard J. Dalcourt

MAGI BOOKS, INC.

33 BUCKINGHAM DR.   ALBANY, NEW YORK 12208

Manufactured in the United States of America by
the Hamilton Printing Co.

BOOKS BY YVES SIMON

CONTENTS

VII TIME

The Aristotelian discussion of time is basic, 129. Analysis of Aristotle's procedure, 129. The problem of the reality of time, 131. Time is a being of reason, 131. Time not identical with motion, 132. Time as duration, 132. Duration with and without motion, 133. Activity and duration without motion are qualitatively supreme but brief, 133. Notion of duration is analogous, 134. Usefulness of poets on the subject of time, 136. Newtonian and Kantian time a simulacrum, 136.

VIII PHILOSOPHERS AND FACTS

I. *The Notion of Empirical Science.* Ambiguity of the terms "empirical science" and "deductive science," 139. The points of view from which the various sciences may be said to be rational or empirical, 139. Since all sciences are based on facts, the problem is to determine the nature of the facts on which philosophy is founded, 144. II. *The Notion of Fact.* Facts are empirical absolutes, 144. The differentiation of the various sorts of facts, 145. It is impossible to incorporate into philosophy a scientific fact as such, 149. III. *The Myth of a Philosophy Based on Scientific Facts.* Scientism is rife among scientists, 149. It has also been adopted by various philosophers, 150. As Bergson has shown, it is sophistic, 151. IV. *Common, Scientific and Philosophical Experience.* All the fundamental philosophical facts are derived from common experience, 153. But many vulgar facts are not philosophical ones, 154. Philosophical facts may also be derived from scientific experience, 155. The establishment of moral facts, 158.

IX SCIENCE, SCIENTISM AND REALISM

I. *Epistemological Pluralism.* Scientism and its opponents, 163. Philosophical disciplines are the most truly scientific, 165. Science has evolved pluralistically, 166. The reaction of various Thomists, 167. Scientism the counterpart of ontological integralism, 169. Pluralism justified by the analogy of being, 169. II. *The Reality of Scientific Objects.* Some apologists make use of a faulty epistemology, 170.

EDITOR'S PREFACE

Yves Simon was born in Cherbourg in 1903. Despite a child-
hood illness which left him lame, he pursued his studies vigor-
ously. After attending the Lycée Louis Le Grand in Paris, he
did graduate work at the University of Paris where, besides taking his licenciate in letters, he also received a diploma of higher
studies in philosophy and a certificate of studies in the sciences.
After four years in medical school he decided to dedicate himself totally to philosophy and so went on to take his licenciate
and then his doctorate in philosophy at the Catholic Institute in
Paris.

After teaching for eight years at the Catholic University of
Lille, he was invited in 1938 to the University of Notre Dame,
where he stayed for ten years. He then joined the Committee of
Social Thought of the University of Chicago Graduate School,
where he taught until 1959 when illness forced his retirement.
During his whole teaching career he was also in great demand
as a lecturer; the reasons why this was so are well exemplified by
his Aquinas Lecture at Marquette on *The Nature and Functions of Authority* and the Charles Walgreen Lectures at the
University of Chicago, entitled *Philosophy of Democratic
Government*. They manifest the same qualities which made him
one of the great teachers of our time: depth and clarity of vision
combined with a flashing lucidity. He died in 1961.

Simon's voluminous output deals mainly with three areas:
political philosophy, metaphysics and ethics. Especially note-
worthy are his *Introduction à l'ontologie du connaître, Critique*

de la connaissance morale, *The Road to Vichy: 1918–1938*, *The March to Liberation*, *Prévoir et savoir: études sur l'idée de la nécessité dans la pensée scientifique et en philosophie*, *Traité du libre arbitre* and *Philosophy of Democratic Government*. Besides these, special mention must be made of his translation, with John Glanville and Donald Hollenhorst, of the monumental *The Material Logic of John of St. Thomas*. Simon's writings brought him international renown and made him, in company with Maritain and Gilson, one of the recognized leaders of Thomistic philosophers. What perhaps above all distinguished him as a philosopher was his ability to relate in a most illuminating way the concrete problems of everyday life and philosophical theorizing.

At the time of this death Simon was working on a variety of philosophical projects. On the basis of what he left, various editors have been preparing a number of publications. The present volume appears after the following posthumous works: *A General Theory of Authority* (1962), *The Tradition of Natural Law* (1965), *Freedom and Community* (1968) and *Freedom of Choice* (1969). Several more are in preparation. Of interest too are the biography and complete bibliography expected to be available shortly.

* * *

The papers presented here come from several sources. The first seven are based on a series of lectures which Simon started to give in 1959 at the University of Chicago as a member of the Committee on Social Thought, but which he could neither finish nor revise because of his terminal illness. The eighth and ninth chapters are translations of articles published much earlier in European journals of philosophy. The concluding chapter is an abridgement of the translation and revision Simon himself made of the first chapter of his *Prévoir et savoir*.

Despite the disparity of their origin these papers tie in well together and form a unity of classical and contemporary insights that should be of some value to the general reader and also, in some instances, to professional philosophers. Simon had studied Aristotle intensively but had a broad scientific and medical background too, besides being familiar with modern philosophy. Listening with a sympathetic ear to the various traditions within these areas, he was often able to make clear what in Aristotle is no longer tenable but also what are the many notions and theories of permanent value. Consequently, he believed that very often the discoveries of the modern world could be best understood and interpreted within an Aristotelian framework. In the spirit of free inquiry which was so characteristic of him, I invite you the reader to continue on and decide for yourself whether or not he was right.

In editing and translating these papers my concern has been to maintain as much as possible Simon's language and style, with their distinctive pungency and directness, while giving a clear and idiomatic rendering. In this I was aided to some extent by having at my disposal most of Simon's manuscript notes. For, he had accumulated numerous folders of unpublished original materials, all of which are being kept at Notre Dame's Jacques Maritain Center for the convenience of whomever may wish to study them. In the spring of 1967 I had occasion to consult these folders in their entirety and then Prof. Simon's widow provided me with copies of the more pertinent material related to the text. I also had the tape recordings and transcriptions, made by Mr. Richard Marco Blow, of the lectures on which the first seven chapters are based. As editor, I have excised a certain amount of material that was repetitious or not directly relevant and have shifted some parts for the sake both of unity and of continuity. In all this however I made no changes or additions which Simon, in my estimation, would not have approved of. My additions consist for the most part of the notes, but even some of these I drew directly from Simon's text, and the "Selected Readings" at the end of each chapter.

Gerard J. Dalcourt
Seton Hall University

ACKNOWLEDGMENTS

I wish to extend my thanks to Professor Simon's widow, Mrs. Paule Yves Simon, for giving me the opportunity to prepare these papers for publication. I am also grateful to her for reading the text and making many a helpful suggestion on how to improve it.

Special thanks are due to Prof. Arthur Hyman of Yeshiva University who went out of his way to clarify for me some points concerning the work of Maimonides.

I have also been most lucky and thankful to have the help of Sister Luke McSweeney who typed and retyped the manuscript.

To my wife Catherine, whose kind forbearance of the many hours I spent in my study made possible the completion of this and other projects, I owe much gratitude.

Chapter VIII appeared originally under the title "Les Préoccupations expérimentales des philosophes et la notion de fait philosophique" in the *Revue de Philosophie*, N.S., vol 3 (1932). Permission to use it has been granted by Librairie Marcel Rivière (22 rue Soufflot, Paris 5e), which publishes the new series of this journal.

Chapter IX was first presented orally at the days of study, held at Louvain, September 24–25, 1935, on the theme of the interrelations of philosophy and science. The speeches of this convention together with the complementary discussions were first published in a volume entitled *Philosophie et sciences* by Le Saulchoir (then located at Kain, Belgium) in 1935. The following year they were reprinted, without the discussions, in the *Revue Néoscolastique de Philosophie*, now called *Revue Philosophique de Louvain*. Both publishers have granted permission to translate Simon's article.

The original version of the last chapter first appeared under the title "Travaux d'approche pour une théorie du déterminisme" in the *Annales de l'Ecole Des Hautes Etudes De Gand: Etudes philosophiques,* III (1939), pp. 189–236. This was expanded to form the first chapter, entitled "La Théorie du déterminisme," of Simon's *Prévoir et savoir,* published in Montreal by Editions de l'Arbre in 1944.

CHAPTER I

THE GREAT DIALOGUE OF
NATURE AND SPACE

In the history of ideas the two words "nature" and "space" immediately bring to mind two characters in a great dialogue.

The first is Aristotle. Historically there is such a thing as the philosophy of nature. It is not particularly daring to say that the archetype of the philosophic endeavor to understand nature, of the endeavor to interpret nature philosophically, is the work of Aristotle. Indeed, what is central in his thought is the idea of nature—an idea that we have come to understand better and better thanks to the work that has been done this last generation in the study of the pre-Socratics. For instance, there is a young scholar who has published a dissertation which is a part of a long and complex inquiry into the meaning of nature in early Greek philosophy. His name is Charles Kahn; his book, *Anaximander and the Origins of Greek Cosmology*.[1] I expect a great deal of such studies for a better understanding of Aristotle. Then too, the last forty years have seen a good deal of investigation into the development of the thought of Aristotle.[2] When I was a young student it was still the popular opinion that Aristotle had written all his works in the same period of development— the period of his complete maturity. Ever since Jaeger,[3] scholars have given more and more attention to the chronology of the writings of Aristotle. Jaeger started with the *Metaphysics*. He never persuaded anybody that his chronology was right, but he did convince everybody that the chronological consideration was relevant. A number of his followers attempted to divide the

1

treatise *On the Soul* and other Aristotelian works into fragments from various periods. My impression is that there is a
great deal of arbitrariness in any such determination of chronology. It depends very much on what the writer considers genu-
inely Aristotelian or Platonic or of some other provenance. But
the result which is here to stay is the realization that in what we
possess of Aristotle's writings, there are the obvious and certain
marks of a development; that some passages belong to much
earlier and others to much later periods; and that a number of
difficulties could be solved if we succeed in dating every significant passage. Unfortunately, to the best of my knowledge our
success in this dating has been rather limited.

There is another point of great significance for the understanding of Aristotle's philosophy of nature, that I personally
became aware of only recently, although I should have realized
it long ago. It is clear that the Aristotelian corpus, the body of
Aristotle's writings that we are familiar with and that was also
known to the Middle Ages and to the Renaissance, apparently
was not read much by the philosophers who lived in the couple
of centuries after Aristotle.[4] Thus, the Stoics did not know the
Aristotelian treatises which we know. But what is very significant is that though they did not know the Aristotle that we
know, they knew another one whom we do not know except
through a few fragments and a few reports and allusions. They
knew the Aristotle of earlier works, of the dialogues, on which
people like Jaeger have worked so hard. Another book came
from Belgium recently that tells how scholars are looking here
and there for fragments of these lost works.[5] It would be wonderful if one day an excavation would uncover complete manuscripts of them! Since excavations have yielded so much in the
last few years, it is permissible to hope that eventually we shall
find this unknown Aristotle which was the Aristotle known to
the late Greek philosophers. What is especially interesting is
that the Stoic philosophy of nature is suspected to be a development of this earlier Aristotle. This is fascinating, for the later
Aristotle is not completely independent of the earlier Aristotle.
So if we had a hint at what the earlier Aristotle was, by com-

paring that with the later Aristotle we might learn a few things of great interest, in particular concerning the subject of nature. In Stoicism—this is of tremendous importance—the universe has the unity of an individual. But it has the unity of an individual living being, not that of a work of art. It is not like a temple or a building; it is similar rather to a big elephant, that is bigger than any other. But in spite of its all-inclusiveness, it has a unity of individuality and is animated. Is such a view Aristotelian? It certainly is not compatible with his more mature works, with which we are familiar. Now, if we are warranted in suspecting that it was the line of Aristotle twenty years earlier, then some features of the late and mature Aristotelian philosophy of nature might be interpreted as remnants of an earlier period.[6] It is through such trials and errors and such very uncertain working hypotheses that we progress in the understanding of these difficult subjects.

We are discussing nature and space. But if Aristotle is typically a philosopher of nature, who is the philosopher of space *par excellence?* Who is that least inhibited of human minds who interpreted nature, from A to Z, not in terms of natures [7] but in terms of space? It was of course Descartes. For Descartes, extended substance, which could just as well be called extension, is without doubt simply space.[8] Furthermore, the extended substance of Descartes, or space, is not a nature but something quite different. Unfortunately, this is again a point that I came to realize only rather recently. Fifteen years ago I would have told you that Descartes simplified the situation a great deal by retaining only two natures instead of the indefinitely many natures which made up the Aristotelian universe. I would have said that there were only two natures in Descartes, the extended substance and the thinking substance. But I think it is incomparably more accurate to say that there are no natures in Descartes. The extended substance is not a nature in any sense.[9] It is Archimedean space; it is the non-qualitative space of the Greek geometricians. As to the thinking substance, it is not a nature either, but something else, because for Descartes thought is characterized by pure passivity; ultimately it is God acting on the thinking substance that makes it have this or that idea. That there are no natures in Descartes is quite consistent with some fundamental Cartesian ideas. Why is God so important in the physics of Descartes? It is because space not being a nature, and extended substances being nothing but extension or space, the regularity of nature's laws is not guaranteed, as it would be in Aristotle, by the natural constitution of objects. In order that physics be possible, Descartes needs divine immutability. For him, the world does not consist of things each of them having its own whatness and its necessary ways of acting and reacting. Rather it is made of a thing which is not a nature and which has no nature. How then do we know that it is going to behave in determinate fashion, so as to make a scientific knowledge of it possible? Here, Descartes brings in a divine promise of stability. God, being immutable and having decided once and for all that the laws of motion would be such and such, fear not, the laws of motion will not change. In regard to this latter point, there is no doubt whatsoever concerning the accuracy of this interpretation of Descartes.[10]

The great characters in the dialogue we are considering are Aristotle and Descartes. Now, are they opposed the way we might set in opposition the philosophic and the scientific interpretation of nature? Factually, they probably are; in intention, certainly not. Descartes, and Galileo as well, do not propose to replace the old philosophy of nature by a science of nature that would not be philosophic. Indeed, identifying science and philosophy of nature and being certain that what they offer is the true science, they hold that it is also and necessarily the true philosophy of nature. Interpretations in terms of the hypothetico-deductive method, in terms of the "saving the appearances" approach, in terms of geometrical constructs which account for phenomena but do not necessarily express the truth about nature, all these are totally foreign to the founders of modern science. Galileo, and still more Descartes, are sure that their science is the true philosophic understanding of nature. These men, who very certainly did not distinguish between the philosophic and the scientific interpretation of nature, may well be historically at the origin of a development which we may describe as a non-philosophic though scientific interpretation of nature. This may be the case in fact, but it was certainly not their intent. They were without doubt aware of the novelty of their science of nature and they appreciated its newness. But they did not wish to give it an epistemological character by which it would be distinguished from a philosophic approach to nature.

Because of Descartes' strenuous opposition to the Aristotelian approach, it might seem that he rejected any philosophy of nature. But this is not so. On this subject you can consult a really great historian of science, Pierre Duhem, whose major work is entitled *The System of the World*.[11] It is a little obsolete now because he died in 1916 and much work has been done since then. It has been corrected on various points by recent investigation but will remain a classic for a few more centuries. It is a tremendous mine of information, but necessarily long. I recommend to you as most enjoyable and stimulating reading, a shorter opus of his. I am speaking of a short book of about one hundred and forty pages entitled *Sōzein ta Phainomena,* which means *Saving the Appearances*.[12] It is a study on the notion of physical theory from Plato to Galileo inclusive. That Greek phrase comes from Simplicius, a fifth century neo-Platonic commentator of Aristotle. In an illustrious passage of his commentary on Aristotle's *On the Heavens (De Caelo)*, Simplicius says that for Plato the purpose of a physical theory is to save the phenomena, to safeguard the appearances.[13] To illustrate what he meant consider how the shadows in a room seem to move. Hour by hour they seem to slowly shift about. We may imagine that the sun has been turning around the room or that the room has been turning around the sun. In both cases the appearances are saved and the problem is to find out what is the more satisfactory way to save the appearances. When we have found it, shall we say that our geometrical constructions concerning the sun, the earth, the planets and the stars express the truth about nature? Or shall we say that they are very handy since they give us a unified vision of nature and make for the prediction of astronomical and some other events but that yet, perhaps in the next generation, a smarter person may find a better way of saving the appearances? Success in saving the appearances does not exclude the possibility of a still more successful pattern. You find this approach in Plato.[14] Is it rejected by Aristotle? I do not think so, but that is certainly not what he himself does. He intends to explain nature, to say what is purely and simply true about nature.

The great merit of Duhem is that he has a unique sense for finding the relevant texts in the most unexpected places. In this book he also reports the views of Poseidonius, a representative of Middle Stoicism who flourished in the first century before Christ. I would not be particularly interested in him if Duhem had not translated and quoted a long page containing his ideas about astronomical theories which save the appearances but do not explain the reality of nature.[15] Duhem also points out that we find the same interpretation in an unexpected connection in a very great thinker of the twelfth century, Moses Maimonides. We could call him the Jewish Aquinas. Of course, every resemblance being mutual, we could just as well say that Aquinas is the Christian Maimonides. At any rate, in his *Guide of the Perplexed* Maimonides holds to the "saving the appearances" interpretation of astronomy for a very special reason. Like Aristotle, he believes if not in the divinity, at least in the very lofty spirituality of celestial bodies. Because they are so superior, they are beyond the range of our understanding.

Being unable to understand and explain these exceedingly lofty substances, all we can do is to place behind the appearances geometrical constructs which are very useful for purposes of organization, simplification and prediction, but which do not tell us what those lofty beings are. He had in mind, of course, mostly the system of Ptolemy, the Egyptian astronomer of the second century of the Christian era.[16]

We meet up now with a very fascinating event in the history of the notion of physical theory. In Maimonides, the "saving the appearances" theory is connected with a mythological view concerning the loftiness of the celestial bodies. Now, against this theory you have Averroes and his school. Averroes was a twelfth century Spanish Moslem and the staunch enemy of the "saving the appearances" method. For him, astronomy has to be philosophic or disappear. It has to say the truth about the nature of celestial bodies or say nothing. Averroes, as quoted by Duhem, said that he was too old to write a philosophic astronomy but that it should be done by somebody. And it was indeed done, by a disciple of his, Al Bitrogi. What happened is well known. The commentaries of Averroes on Aristotle, along with various other Arabian and Jewish works, were received in the Latin West, at the very latest, in the very early thirteenth century. It is an extraordinary story. There must have been quite a good period of friendly cooperation between Moslems and Jews and Christians in Spain in the twelfth century. Without friendly cooperation and all the translations which it made possible, there would have been no Albert the Great, there would have been no Thomas Aquinas, there would have been no Latin Aristotelianism. But as a further result of these translations Averroism became widespread. That it could so quickly become a really active movement and then maintain itself as a philosophic power is quite surprising. For, as soon as it entered the Latin West it ran into the opposition of such formidable men as Albert the Great and Thomas Aquinas, and yet it kept doing quite well until late in the seventeenth century.[17] Its decline, we may note, paralleled the rise of modern physics which destroyed its foundations. I think that political factors also had a lot to do with the survival of Averroism over the centuries.[18] Its survival is even more paradoxical if you consider that it maintained itself in the most different and conflicting ways. Often it was associated with free thinking, disbelief and the opposition of philosophy to religion, while elsewhere practiced and taught by men who were religious themselves. A very strange story and an extremely important one. For centuries men used the philosophical astronomy of this friend and student of Averroes. I think it was taught down to the seventeenth century in some schools. Duhem says that its failure to work out a philosophy of celestial motions was obvious from the beginning. No matter how obvious the failure, the attempt was so fascinating that it attracted attention and dedication for centuries.

All these last remarks were made just to locate Galileo and Descartes in the dialogue we are considering. Do they stand with the "saving the appearances" method, or do they opt for the view that they can explain nature philosophically? So far as their own work is concerned, neither want to have anything to do with the "saving the appearances" interpretation of physical theories. That is splendidly shown by Duhem.[19] I am not too well acquainted with Galileo, but I do know Descartes a little and I can guarantee that Duhem is correct on this point.

As far as epistemological purpose is concerned then, Galileo and Descartes are both similar to Al Bitrogi inasmuch as none of them want to use the "saving the appearances" approach, and they all consciously reject its appropriateness. There is in Descartes a passage [20] on the Copernican system where he seems to speak the language of the "saving the appearances" theory, and because of that passage historians raise the question of his sincerity. There is even a famous paper on the sincerity of Descartes which is based mainly on this passage.[21] It seems obvious though that Descartes is hedging here. He is being diplomatic because he wanted his book, the *Principles of Philosophy,* in which this passage occurs, to be accepted as a textbook in the schools and to quickly displace those in use, which presented so-called Aristotelian philosophies of nature. I say "so-called," because little was left of genuine Aristotelianism in textbooks of the seventeenth century and I would not trust them.

In regards to Galileo, Duhem also has several very interesting works pertaining to his trial. As far as I understand the problem, it was essentially a theological error; not a scientific error of the theologians, but a theological error of the theologians. In respect to the passage of the Bible where Joshua is said to have stopped the sun, the question arises: Does Holy Scripture speak the language of sense appearances or the language of the absolute reality of things? Not so long after the trial of Galileo there was a consensus of theologians that since the Bible was not written to teach astronomy, it could hardly speak any other language than that of sense appearances and popular representation of the relation between the sun and the earth. So, what strikes me as regrettable at the trial of Galileo is not so much a conflict of the Bible with developing science as a failure within theology. For, at the time when the popularization of the Copernican system by Galileo began to cause difficulties, Robert Cardinal Bellarmine wrote to a friend and sponsor of Galileo a letter, which Duhem did not miss the chance to quote, in which he said there would not be any problem if Galileo would be willing to characterize the Copernican system as a hypothesis in the sense of Simplicius, something designed to save the phenomena.[22] But Galileo did not want to buy peace at that price because he did not believe that such was the case. He was so dead sure that it was not a hypothesis designed to save the phenomena that when offered peace at that price he said, "No."

In order to understand a bit more clearly the nature of the problems with which we are concerned, let us reflect a moment upon the connections between the notions of place, of order and of a good state of affairs. I call your attention to a couple of points which are really very familiar. First, think of order. This immediately brings to mind things that are placed in a certain way. Does not order mean primarily a certain way for the parts of a multitude to be placed? If the members of a set are so placed that beginning with an initial one we can see each of them is on the right-hand side of the one before and on the left-hand side of the one after, then we have a totally ordered set. If only some members of the set can be placed in such a definite fashion, then twentieth century mathematicians call it a P.O.S., which means a "partially ordered set." The second point on which I invite you to reflect is this: when we speak of order (and again our primary vision of order is relative to place), we generally have in mind a good state of affairs. The notion of order is heavily loaded with teleology, finality. It takes an additional effort of abstraction to conceive an order without finality. In the real world, things are in order, that is, they are ordered to each other: each of them is in the right place. And thus, order, place, goodness and desirability all involve each other.

There is one scientific discipline which effects that additional abstraction which brings forth a notion of order that does not involve finality. There is one scientific domain in which we unmistakably and patently find patterns of order without any relation to right or wrong, to good or evil, to success or failure. It is, obviously, mathematics. The mathematician works out a notion of order which is completely disengaged from finality. He is not the only one who does it. The logician in his own way does it also. But do mathematicians and logicians deal with the real world? Do they deal with nature or with something else? We are here, I think, right at the center of the problem, or of the family of problems, entailed by the topics of motion, place and time. For if the major protagonists in the dialogue over nature and space are Aristotle and Descartes, it is because the former explains the world in terms of natures while the latter does so in terms of an extended thing which is not a nature but the nonteleological and nonqualitative space of the mathematicians. Always keep in mind this statement of Descartes: "I do not accept or desire any other principle in physics than in geometry or abstract mathematics, because all the phenomena of nature may be explained by their means, and sure demonstration given of them." [23] But in mathematics we have something very extraordinary. We are dealing with a universe of which we would not dare say that it is purely and simply unreal. Mathematical objects do have something in common with reality. However, as a result of our abstractive activity, they lack the very qualities that characterize real things: substantiality, individuality and mutability. And in Descartes you have not only the substitution of space for nature but the most daring mathematization of nature ever attempted.

It would be useful at this point, I believe, to remind you how Aristotle distinguished the various sciences. First, he breaks them down into the theoretical and the practical. Then he differentiates, among the theoretical sciences, the physical, the mathematical and the theological. For Aristotle, the object of the physical sciences is things that are mutable and that are besides perceivable by the senses. We may note that here, with the relation between mutability and perceptibility, arises one of the extreme difficulties and perhaps too one of the great weaknesses of Aristotelianism. You find in Aristotle and in all his commentators a certain willingness to assume that what is perceptible to the senses is mutable, and vice versa. The former is clear enough: the latter, that whatever is mutable is perceptible to the senses, is not.[24] For it does not seem compatible with the presence in nature of things like magnets. Place a certain kind of stone near a piece of iron. Suddenly the piece of iron jumps though you neither see nor feel anything else happening. On the other hand, if a stone which was cold becomes hot when you put it in boiling water, you can feel that the water is hot so there is nothing mysterious about the stone becoming hot. But what about the iron which is attracted by a lodestone? You perceive absolutely nothing related to its sudden jump.

Whatever may be this very annoying and unceasingly problematic relation of the mutable to the sensible (taken in Aristotle's sense of perceivable by the senses), there is no doubt that the object of the philosophy of nature is mutable being. That is where the philosophy of reality starts, with the study of real, mutable beings. It is not first philosophy taken as a synonym of metaphysics, but it is the first to be formulated by men.

Now, here you must be on your guard against a confusion which has been going on for many centuries and which is still very widespread. A few years ago talking with Professor Koyré from Princeton, I told him, "Koyré, I want my students to write an essay on mathematics conceived as a natural science. What in your opinion are the main readings they should do?" We quickly agreed that Comte would be a good example and that John Stuart Mill would be perfect. We both forgot a man greater than Comte and greater than John Stuart Mill; we really should have included that egregious savant, Antoine Cournot, who had the theory that mathematics is both a system of abstract thoughts and an interpretation of the quantitative aspects of the physical world. We kept talking and I mentioned

to him that recently I had read a statement by Northrop [25] to
the effect that for the Greeks, mathematics was a natural science. And again we agreed. But for which Greeks? If we speak
of Aristotle, it is obviously false, and if we speak of Plato, still
more so. It is not true that for the Greeks mathematics was a
natural science. In Plato, in the Plato of the *Timaeus,* the question is very complex because of his particular notion of physical
theories. Remember what Simplicius said, that for Plato physical theories are hypotheses designed to save the appearances.
Those hypotheses are geometrical constructions. Now, they are
not supposed to express physical reality but to save the phenomena. In Aristotle it is a lot clearer. There is a legend according to which Aristotle knew nothing about mathematics. He
certainly was not a creative mathematician. It has been remarked that there are only two philosophers that are great
mathematicians, Descartes and Leibniz. There is a third one who
without being a creative mathematician knows very well what
mathematics is all about. It is Plato. But they are all. Spinoza
was in good faith when he said philosophers proceed according
to the method of the geometricians and then, as to be expected,
declared that Aristotle knew nothing about mathematics. That
Aristotle is not a mathematician by inclination is obvious—he
is a naturalist, definitely. But, consider what he has written about
mathematics. The assignment is easy because the texts have
been collected by the illustrious Dr. Heath, the great master of
Greek mathematics, who wrote a four volume commentary on
Euclid and a textbook of Greek mathematics. After he died his
widow was delighted to see that he had also put together all
the texts of Aristotle on the subject of mathematics, with some
very interesting commentary.[26] If only you study that collection
of texts you will see that Aristotle has a very sharp sense for
mathematical abstraction. Every sort of knowledge is abstract in
some way, but between the physical and the mathematical there
is an essential difference in abstraction. It is a difference which
we can appreciate if we consider the two meanings of the
word triangle. That word can designate a gadget which you can
buy in an office supply shop and with which you can draw

ninety and forty five degree angles. But you can also designate by triangle a geometrical figure. I can draw one on the blackboard, but of course you know that what is on the blackboard is not the geometrical triangle; it is something physical, an area of particles left by a bit of chalk on a background of slate. A mathematical triangle would not catch fire in a house that burns down. But if I want to change a mathematical entity, even one as close to physical reality as a triangle or a sphere, into a real physical object, as I achieve my purpose I deprive it of its mathematical condition and it assumes a physical condition. That is clear in Aristotle. Thus, this is a very old story. Mathematical entities are not real or physical ones. However, the illusion that mathematics is a science of real quantity kept haunting the human mind down to the great scientific revolution of non-Euclidean geometries that took place in the last century. It would not have shocked people out of their wits if the nature of mathematical abstraction had been borne in mind; but in spite of the satisfactory clarification of the question very early in Greek science and philosophy, for a lot of obvious reasons, the dream of a mathematics that would be a science of the quantitative properties of the real world, in other words, a mathematics that would be a physics of quantity, that dream kept haunting the human mind—and it was shattered violently by the non-Euclidean geometries. The drama, the crisis, the threat of skepticism, the desperation caused by that great revolution, all that, in my opinion, was superfluous and could have been avoided if more attention had been given to what has been known for many centuries concerning the characteristics of mathematical abstraction: it implies features which render the object incapable of physical existence so that the conditions of physical existence are always and from the very beginning out of the picture as soon as abstraction assumes a mathematical character. This is of the greatest importance for all that is coming.

By comparing Descartes with Aristotle we have already perceived the differences between one physicist, Aristotle, who means to deal with physical reality, and another physicist,

Descartes, who likes to assume that there is no more in the physical world than in the world of the mathematician. That is really what Descartes means in his treatise *On the World,* when he says that philosophers distinguish several kinds of motion and quotes in Latin the definition of motion by Aristotle, declaring that it makes absolutely no sense and that it is not even translatable—although it had already been translated from Greek into Latin. (The humanists were a lot better in Latin than in Greek. Do not forget that). Moreover, he says, he is not going to consider in nature any other motion than that which is considered by geometricians as clearer, simpler and more primitive than such clear, simple and primitive notions as those of line and surface. He recalls that geometricians define a line by the motion of a point and a surface by the motion of a line. That cannot mean anything else than this: here is a physicist who, indeed, is a mathematician of genius and who likes to postulate that the components of the physical world boil down to mathematical entities.[27]

If you had asked me only a few years ago whether Descartes is really the founder of modern idealism, I would have said, "Yes." And if you had asked me precisely why I would have answered, "Because in Descartes that which is known first is not the thing but the idea of the thing. That which is known primarily is not independent reality but a mode of the psyche, a mode of my consciousness, a psychological disposition. Then from the characteristics of that psychological disposition, I can infer, through an application of the principle of causality, whether there corresponds something to it as real and how far it does. For instance, if the psychological mode is what I experience when I say it is blue, green, hot or cold, nothing real corresponds to those modes, because they are only confused ideas. But if I speak of extension, number or motion, then we have here clear and distinct ideas and because of their clearness and distinctness I can infer that they really represent something." You can see that all the principles of modern idealism are here.

To elucidate why this is so, let me tell you of a sweet experience in my early youth as a teacher. I think it was during my second or at the latest my third year as a university teacher. One of my students returned to me a paper which was atrociously confused but because of one sentence I gave him at least a passing grade. The sentence was: "With Descartes the relation between knowledge and reality is no longer one of intentionality but one of causality." You can imagine how happy I was, since I have not yet forgotten it after twenty five years. But what does this contrast between intentionality and causality mean? Aristotle, in many places and especially in those great venerable texts of Book Three of the treatise *On the Soul*, says that the soul is in a way all things, that the soul is all things according to sense and according to intellect, that the intellect in act is one with the intelligible in act. The meaning of these texts has been worked out by generations of commentators, the most important for me being Averroes and Aquinas, and the meaning is clear. The meaning is that there are for things two ways of existing. One and the same thing exists in a certain way in nature and in another way in the soul. And its existence in the soul is one with the act of cognition. That is what those texts of Aristotle mean.[28] Whereas for Descartes, in the soul you do not have a thing, you have an idea of it. What is decisive here is this, that this idea of the thing is known first, as if, not having had the pleasure of meeting you, I have seen your photograph and from the features of the photograph I infer what you are in reality. The relationship is causal here. Not one of intentionality but one of causality, as my confused pupil said in the one sentence of his paper that was not confused. That which is understood, in Descartes, is a psychological disposition, an idea. What causes us to rely on that psychological disposition, that idea? Provided you master Descartes' four rules of the method, you can differentiate between ideas that are dependable and those that are not. The ideas that are dependable are so because they are really occasioned by things. The ideas that are not dependable are not occasioned by things. This Cartesian principle, that what we know primarily are not things but rather our ideas of them, is a fundamental principle of idealism.

In the last week or so, however, reflecting on the meaning of motion in Descartes, I have found what seems to me another and at least equally important source of idealism in him and in later philosophers. That reduction of the physical to the mathematical, that bold and so self-assured assertion that motion is nothing else than what geometers mean when they say that a line is generated by the motion of a point, that reduction of the physical world to a thing that has already been treated by abstraction, this is what Descartes does and I think we have here a principle of idealism at least as powerful as his theory of ideas. Descartes is the theoretician who resolves the physical world into a system of objects already treated by mathematical abstraction. He is the philosopher for whom the physical world is made of things already treated by the abstractive power of the mathematical mind.

Descartes, as we have seen, wanted to replace the false old philosophy of nature by a true one. Now, having done it in his own way, what has he actually done? Has he worked out a philosophy of nature, or has he developed an interpretation which if accepted would rule out forever the possibility of a philosophy of nature? The latter, most certainly. That is confirmed in a hundred ways by the evolution of ideas since Descartes.

Here, I should point out that in many cases, perhaps to some extent in all cases, there is some discrepancy between the intentions of a philosopher and the intentions of his philosophy. And, you cannot expect one man, who died at the age of fifty four, who was so prodigiously an innovator and, accordingly, a solitary—you cannot expect him to achieve full awareness of what he is doing. That happens to nobody. It most particularly cannot be expected under those circumstances. However, we should concentrate here on the change in the notion of nature. I have already told you there are no natures in Descartes. Is there such a thing as Nature? After what we have seen about his mathematization of physical things, it can be strongly suspected that if the word nature can still be used in his philosophy it is in a different sense, a sense patterned by this mathematical interpretation.

If we compare Aristotle's philosophy of nature with Cartesian physics, we have two conceivable, and very probably factual, divergencies. The first concerns the content of the science of nature. Aristotle interprets the physical world in terms of natures; Descartes, in terms of space. Thus, in regard to content there is a first discrepancy. The second concerns the epistemological form. Descartes does not—any more than Galileo—intend to substitute a new science of nature for the old philosophy of nature. They both are sure that what they are proposing is the true philosophy of nature in opposition to a false one. That is clear. Now, considering not what they wanted to do but what history, the subsequent development of ideas, shows that they have done, we may very well conclude that they have in fact substituted a nonphilosophic for a philosophic interpretation of nature. In other words, ever since the positivistic awareness made it possible to state problems more clearly in the last 150 years there has arisen the problem of the very possibility of a philosophy of nature. And that problem has metaphysical consequences. I used to have a friend who was, and probably still is, quite a good medievalist, who one day explained to me that the bad thing with Thomistic metaphysics is that it implies the possibility of a philosophy of nature. True, it does. Now, he himself considered that the possibility of a philosophy of nature had been destroyed by the work of the great thinkers of the seventeenth century, above all, Galileo and Descartes. Thus if there were to be any future for metaphysics, metaphysics would have to assume such a form as to be possible independently of any philosophy of nature. We are here at the heart of a system of problems of great relevance. These problems concern the content of what the philosophers have to say about nature. They also concern the epistemological question of the very possibility of a philosophy of nature. And if this question is answered in the affirmative, if a philosophy of nature is possible at all, then, you can see that in our progress in the philosophic interpretation of nature we shall have constantly to carry out an epistemological reflection on what we are doing. All that promises to be very difficult, but if we are only half equal to our task it should also be very interesting.

NOTES TO CHAPTER ONE

1. Charles Kahn *Anaximander and the Origins of Greek Cosmology* (New York: Columbia University Press, 1960). Kahn taught for several years at Columbia and is now (1968) at the University of Pennsylvania.

2. For a brief but enlightening treatment of this point, see David Ross, "The Development of Aristotle's Thought," *Aristotle and Plato in the Mid-Fourth Century* (Göteborg, 1960), pp. 1–17.

3. Werner W. Jaeger, *Aristotle: Fundamentals of the History of His Development* (2d ed.; Oxford: The Clarendon Press, 1948). Originally published in German in 1923.

4. See Emile Bréhier, *The Hellenic Age* (Chicago: University of Chicago Press, 1963), pp. 232–5. For fuller treatment and references see Joseph Moreau, *Aristote et son école* (Paris: Presses Universitaires de France, 1962), pp. 272–81.

5. Joseph Bidez, *Un Singulier naufrage littéraire dans l'antiquité* (Bruxelles: Office de Publicité, 1943).

6. For a further elaboration of this point, see Chapter Six.

7. For Aristotle the primary sense of "nature" is a thing's essence taken as a source of activity. For example, a man is an individual substance with an essence or set of characteristics which distinguish him from other kinds of substances, such as dogs or trees. As a result of having different essences, these different kinds of substances also tend to act in correspondingly different ways. Hence we say it is man's nature to think but not a dog's. In a broader derived sense "nature" also means the totality of substances taken in their dynamic, "natural" interrelations. The problem that Simon is concerned with here is whether Aristotle is right in saying that the world consists of individual substances acting in accord with their own natures and can be understood only in terms of these natures or whether Descartes was warranted in rejecting such a view.

8. "Space or internal place and the corporeal substance which is contained in it, are not different otherwise than in the mode in which they are conceived by us. For, in truth, the same extension in length, breadth, and depth, which constitutes space, constitutes body; and the difference between them consists only in the fact that in body we consider extension as particular and conceive it to

change just as body changes; in space, on the contrary, we attribute to extension a generic unity, so that after having removed from a certain space the body which occupied it, we do not suppose that we have also removed the extension of that space, because it appears to us that the same extension remains so long as it is of the same magnitude and figure, and preserves the same position in relation to certain other bodies whereby we determine this space." René Descartes, *Principles of Philosophy*, trans. E. Haldane and G. Ross (London: Cambridge University Press, 1931), II, 10.

9. See Etienne Gilson, *The Unity of Philosophical Experience* (New York: Charles Scribner's Sons, 1937), pp. 198–214.

10. *Ibid.*, pp. 203–6.

11. Pierre Duhen, *Le Systéme du monde* (Paris: Hermann, 1913–58), 10 Vols.

12. *Sōzein ta phainomena; essai sur la notion de théorie physique de Platon à Galilée* (Paris: Hermann, 1908).

13. For Simplicius' report on Plato's views, see his *In Aristotelis quatuor libros de Caelo commentaria*, II, nos. 43, 46.

14. *Timaeus*, 28–29.

15. Duhem, *op. cit.*, pp. 9–11.

16. See especially *Guide of the Perplexed*, Part II, chaps. 11 and 24.

17. Very little has been written in English on Averroism. For an accessible, relatively full summary of research findings and the main sources, see "Averroism, Latin," *New Catholic Encyclopedia.*

18. For example, in the early fourteenth century John of Jandun and Marsilius of Padua, both Averroists, opposed the political power of the papacy and were protected by the Emperor Louis of Bavaria.

19. Duhem, *op. cit.*, pp. 125–40.

20. *Principles of Philosophy*, III, 13–19.

21. G. Milhaud, "La Question de la sincérité de Descartes," *Descartes savant* (Paris: Alcan, 1921), pp. 9–24.

22. Duhem, *op. cit.*, pp. 128–9.

23. *Principles of Philosophy*, II, 64.

24. A parenthetic note of importance. Whoever would be willing to devote ten years or so to the study of the relation between the mutable and the perceptible in the philosophical physics of Aristotle would be rendering scholarship a real service. If anyone is tempted to write a doctoral thesis on the subject, he ought rather to choose some other topic in order to get his doctorate within a reasonable time. Then he can take ten years to work on this subject.

25. F.S.C. Northrop, professor at Yale and the author of *The Logic of the Sciences and the Humanities.*

26. Sir Thomas Little Heath, *Mathematics in Aristotle* (Oxford: The Clarendon Press, 1949).

27. *Oeuvres de Descartes.* Edited by C. Adam and P. Tannery (Paris: Cerf, 1897–1957), vol. 11, pp. 39–40.

28. For a fuller explanation, see Y. Simon, "An Essay on Sensation," *Philosophy of Knowledge,* edited by R. Houde and J. Mullally (New York: Lippincott, 1960), pp. 55–95.

Readings (An asterisk indicates that the title is available in a paperback edition)

* R.G. Collingwood, *The Idea of Nature* (New York: Oxford University Press, 1960).

  Pierre Duhem, *To Save the Phenomena: An Essay on the Idea of Physical Theory from Plato to Galileo,* trans. E. Doland and C. Maschler (Chicago: University of Chicago Press, 1969).

* Pierre Duhem, *The Aim and Structure of Physical Theory,* trans. P. Wiener (New York: Atheneum, 1962).

* Etienne Gilson, *The Unity of Philosophical Experience* (New York: Scribner, 1936).

* A. Koyré, *From the Closed World to the Infinite Universe* (New York: Harper, 1957).

  V.E. Smith, *Philosophical Physics* (New York: Harper, 1950).

* A.N. Whitehead, *Science and the Modern World* (New York: Free Press, 1967).

Chapter II

HOW WE EXPLAIN NATURE

Let us now turn to the philosophic problems of change and motion. For the time being, I leave the relation between those two terms indeterminate. In many contexts they can be used synonymously, in others they should not.

To grasp the fundamental meaning and value of the speculation of the philosophers, and to a large extent that of the scientists also, I think that the best thing to do is to reflect on the sentiment of wonder caused by the experience of change. You know the famous words of Aristotle that wondering *(thaumazein, admiratio)*, is the beginning of science.[1] The example he uses is the existence of incommensurable magnitudes. The fact was discovered by the Pythagoreans and it was integrated by Eudoxus. This was one of the great scientific events in the period of the great Greek philosophers. Aristotle remarks that when we trace the diagonal of a square and try to express both the side and the diagonal in integers, we fail. Could we not succeed by using smaller units, for instance, measuring them in millimeters instead of inches? Again we fail—and no matter how small the units we use we shall never find a unit which can be contained an integral number of times both in the side of the square and in the diagonal. We wonder about it and this is the beginning of a science. Aristotle goes on to remark that after you have understood why there is incommensurability between the side and the diagonal of the square then you would again wonder if they were shown to be commensurable. So, science is included between two acts of wondering. What is the act of wondering which is the point of departure of philosophic, and to a large extent of scientific, speculation about change? It is the strange appearance, the puzzling nature, the paradoxical character of the fact of change. If a person is unaware that there is something strange, paradoxical and wonderful about change, you know that he has not yet reached the lower limit, the threshold of philosophic intelligence—philosophy has not yet begun for him. To hold that change is simply a matter of course is unphilosophic. There is really a problem: How is it possible at all that there should be a course of events? The people who say that they consider change a matter of course are not very much aware of what is going on in themselves because, in fact, everyone on occasion expresses bewilderment and even resentment relative to the fact of change.

It is interesting, once in a while, to stop and to consider how art in general and perhaps poetry in particular may serve to introduce the mind to the meaning of philosophic problems. It should definitely be part of the program of any university or college to do things like that, to show how there is such a thing as a poetic introduction to the deepest problems of philosophy. But so far as an introduction is concerned, anyone of a number of approaches can be good. For instance, if you want to have a vivid realization of what is mysterious and paradoxical about this overwhelming fact of change, read some poetry! Read what the poets have to say about aging, about the shortness of human life, about death, about the perpetual alteration of nature, or about the melancholy of every season. Those things are done by poets in all conceivable ways, from the silliest to the loftiest and profoundest. No doubt, the poetic soul of man succeeds in expressing a mystery which like most mysteries is safely hidden in its own familiarity.

You will understand better the mystery of change and how paradoxical and unbelievable we all consider the fact of change if you pay attention to the everlasting answer to the problem of change. This everlasting answer can be phrased in very few words: "Change is more apparent than real. Do not let this problem of change bother you too much. We do not say that there is no change whatsoever or that it is a sheer appearance, but only that there is not so much of it as we might believe at first. In many cases what puzzles us in change pertains to appearances rather than to reality. In many cases we do not have to wonder why a certain subject has become such and such: it already was such and such." Let us be honest and realize that we are every day indulging in explanations of this kind. Sometimes they make sense. Very often they do not. I am thinking in particular of the generation before me when explanations for heredity were extremely popular. They germinated in a few scientific brains and then were popularized by novelists. Their success was fantastic because it gave peace of mind at such a low cost. "Why is he crazy?" "Well, you should have known his grandfather." "Why does she do things like that?" "All of her family are like that." Do you see the essence of this explanation? You wonder about the appearance, for instance, of some unpleasant behavior in a person and the answer is, there has been no emergence, it was already there a generation before and another generation before and perhaps at the time of Julius Caesar it was already there. The emergence, the novelty —that is the proper word—is seeming, it is not real. There is nothing to wonder about: this object has not become such and such; it already was so. Do you see how we cut down to a minimum the reality of change? If so, you have understood one of the privileges of materialistic explanation.

Materialistic and materialism, as I am using the terms here (for we are not in ethics but in natural philosophy), refer to a certain system of causality and of explanation through a certain system of causes. No moral connotion is involved. I take this precaution because we all know in what a loose way the word materialism is used in our time, with connotations of debased and debasing, and if such connotations are not consciously and systematically ruled out everything gets confused. By a materialistic explanation I mean a certain way of explaining through material causes. Is every explanation through material cause *qua* material cause materialistic? Certainly not. What distinguishes the materialistic explanation among the many forms of explanation through material causes we shall see later. The point here is that any moral connotation is left out of the picture.

We already have enough to understand why systematic explanation through material causes is something very attractive, permanently attractive. I do not say that it is wrong, but of course it can be forced. That would be a third definition of materialism. When the use of material cause in explanation is overdone or forced, when there is such use at any cost, then we can speak with propriety of materialistic explanation. Anyway, you can see how and why we are attracted by explanations through material causes. This is the obvious way to reduce to a minimum whatever is puzzling, mysterious and paradoxical in novelty. Novelty disappears as the continuity of the material cause is brought forth.

There is a great scientific investigator who spent his life studying this subject and writing about it. He is Meyerson, a strange and congenial person, born in Poland in 1859. He was at first a chemical engineer—I fear he was not much of a success as an engineer—in central Europe. A few years before the First World War he settled in Paris and began to write in perfectly readable though undistinguished French extremely important books that anyone who wants to know what science really means should read. It is not so easy for philosophers or for men of general culture to find scientists who know what science is all about and know how to express it. He is one of them, very definitely. Indeed, he was a man of fantastic erudition who knew both philosophy and science. Really something very solid. There is one philosopher that he does not understand, Aristotle; but that does not matter because I can help you with Aristotle. The first book of Meyerson was entitled *Identity and Reality*. It is the only one which was translated into English. It develops an idea that he expatiated upon in four or five other thick books without ever becoming repetitious—which is remarkable. The idea is this: The description of science given by the early positivists was quite inadequate and wrong.

The term "early positivism" refers especially to Comte but

would also include Mach. Auguste Comte (1798–1857) was the founder and promoter of the movement he himself called positivism because he based it on the positive sciences. Ernst Mach (1838–1916), an Austrian philosopher and physicist, was the real founder of Vienna positivism. You can find an anticipation of the future alliance of Vienna positivism and American pragmatism in a letter of William James to his wife around 1890 in which he says he spent a delightful day in Vienna with a charming man, Prof. Ernst Mach. Early positivism, represented by Comte and Mach, should be distinguished from twentieth century logical positivism, developed by the followers of Mach who constituted the celebrated Vienna Circle: Neurath, Carnap, Feigl, Hans Hahn and a few others. Mach and Comte really had extremely different minds with widely diverse views on the role of science. One thing, however, they certainly did have in common: for both of them, science is *not* designed to explain reality.

According to Comte, the notion of cause belongs to the earlier metaphysical age of mankind and positive science simply rules it out, replacing it by the notion of law. For instance, you are puzzled by a phenomenon. After a while you succeed in remarking in this phenomenon a particular case of a more general phenomenon. In other words, you detect a particular case of a law, because in Comte a law is nothing else than a general phenomenon. But then, that is all; you are finished; having recognized the more general in a particular event, your job is done. Would you call that explaining the particular event? Comte does, but that is only a question of words. More fundamentally, he rules out any inquiry into causes; he wants scientific curiosity to be satisfied with the recognition of the more general in the more particular, the recognition of a law in the fact, and he wants you to call that an explanation. But nobody takes him seriously on this. Everybody sees that if science is purely legal—here I am deliberately using the word in an unidiomatic sense that is nevertheless easily understood, since this is the German and the French usage—if science is purely legal, that is, if science is concerned purely with laws but not with grounds, not with causes, then this means to all of us that science is not explanatory. Then what is it good for? For Comte, I know what it is good for. It is good for the control of nature, man's control of physical nature through the prediction of occurrences, and much more importantly, science is good inasmuch as it supplies the means of reorganizing society. The founder of positivism is not a scientific investigator; he is a social reformer. He is interested in science because he expects science to supply a new system of dogmas through which society can be unified again. So he is basically a social reformer. He is intensely concerned with unity in society, but he is not interested in the material causes of social forms and events like the Marxists are for instance. His concern is with the mental, the intellectual causes of social forms and processes. If there is a question of unifying society he will not for instance, emphasize the establishment of a unified technological environment; he is concerned rather with ideologies and remarks that since the Middle Ages Western society has been in a turmoil of opinions, hence the impossibility of achieving order, peace or orderly progress. Not trusting religion to supply the dogmas that a new society needed, he expected science to do it. And immediately he was led into an immobilization of his scientific development which soon caused really amusing occurrences. It has been said quite truly, that no man more systematically made himself blind to the reality of science than the founder of positivism. When I speak of those men of science who can tell you what science is all about, I do not include Auguste Comte. He did not know what it is all about. He blinded himself to what was going on. The great revolutions that were taking place in physics he ignored, or criticized. Why? Because if such things are allowed to take place, his dogmas will break down and science will no longer play the part that it is designed to play in the reunification of society. It has been written that modern science would have been impossible if Comte had been taken seriously by physicists. But fortunately he was taken seriously only by physicians, which was inconsequential.

As to Mach, I do not think he has the same social purposes in mind. He is really a pragmatist, a pragmatist of a particular sort. His pragmatism is an intellectual and scientific one. The dominating idea of Mach in his interpretation of science is, I think, very simply this: if we wanted to know without scientific means, we would be overwhelmed by the weight of endless multiplicity. In order to be able to know and to find your way without being crushed by the multiplicity of the facts, we devised a number of simplifications, or shortcuts; we applied constantly the *principle of parsimony*.[2] Parsimony is for Mach the key for the understanding of what science does. Nonscientific ways of reconnoitering the world are exhaustingly heavy. Science does it in parsimonious fashion, sparing much intellectual effort. Such would be the essential service done by science. So, what we are concerned with here is the problem of explanation.

Comte retained the word explanation but it is soon understood that the word no longer has any substance. Explanation through law alone, without cause, is not understood by anybody as an explanation. So if science does not trace consequences to principles, and effects to causes, what is it good for? We have seen what it is good for in the mind of Comte and in the mind of Mach. Meyerson, in *Identity and Reality,* holds that the spontaneous reality of the scientific inquiry in the nineteenth century is constantly at variance with the positivists' understanding of science. For Comte forbad the scientists to indulge in what he called metaphysical curiosity, to look for causes, reasons, grounds and so on. But actually they do nothing else from dawn to dusk, and oftentimes till late at night; they are always looking for explanations, for causal relations, for reasons and for grounds. And what do they do in their effort to explain nature? *Identity and Reality* can be summed up in these words: The whole history of the scientific mind shows that to explain is to identify. *To explain is to identify.* If we are concerned with change, to explain will consist in showing that underneath appearances which involve novelty there is something which remains identical. If we are concerned not with change but with sheer multiplicity, to explain will consist in showing that things are not so diverse as they look, that if you scratch a little you notice that diversity is superficial and that underneath a diversified surface there is a background of homogeneity. Meyerson has thousands of examples, very skillfully arranged to show that this pattern appears and reappears in all directions of scientific effort. Of course, eventually we run into great trouble. That trouble is expressed by the title of the book. Suppose one day we succeed thoroughly in our endeavor to explain the world through the process of identification. What happens then? At the end of the book Meyerson quotes fragments of Parmenides, who is believed by Aristotle to have held that multiplicity and change are appearances and that the real world is one and motionless. This is at least the traditional understanding of Parmenides and that is how Aristotle interprets him.

Now, does the interpretation of Aristotle provide what was certainly the meaning of Parmenides? When Aristotle attributes a certain theory to Parmenides or to Plato, are we entitled to say, "He should know what was the real meaning of his teacher Plato, and of his ancestor Parmenides. Being closer to them and quite an intelligent person, he should know their systems?" I would not go that far because something which happens to all of us may have happened to Aristotle. He was surrounded by Platonists. Moreover, among the men he knew there may well have been some who drew their inspiration from Parmenides. It is possible that those men to whom he spoke understood Plato or Parmenides in a sense which was not what Plato or Parmenides really meant, yet did exist and had to be discussed. So, I would not say *a priori* that an interpretation of Plato by Aristotle certainly delivers to us the genuine Plato. It may not be the deepest aspect of Plato; it may be a less deep aspect of Plato picked up by more or less mediocre disciples and spread all over. This sort of thing happens all the time. Take for instance a great philosopher who flourished just five decades ago, Bergson. I have known a Bergsonian environment; it was filled with things that Bergson would not have liked very much. But there it was. Those things, whether they express the deepest intention of Bergsonism or perhaps something that occurred incidentally in the development of Bergson, or perhaps something which was inevitable in Bergsonism though uncongenial to the great teacher, did crop out. So, that version of Bergsonism which passes itself off as Bergsonism, is called Bergsonism by everybody and has to be discussed. The same may have happened when Aristotle speaks of Plato or of Parmenides. Sometimes, especially with extremely complex thinkers, the most complex of all being Plato of course, there can be a certain trend that a man of genius represses, which comes up with the disciples. Perhaps it was, indeed, in the man of genius whether he liked it or not. Twenty years later it is a fact that the disciples attribute it to their master. You may quote texts of the great teacher against it; nevertheless, there it is. That may happen. But let us close this digression.

At the beginning of the *Physics* Aristotle discusses the problem of plurality in the physical world and of course he refers to Parmenides. He bears in mind this traditional picture of Parmenides as the philosopher who taught that diversity and change are appearances, that the real world is made of a unit, which single unit is not of a metaphysical but definitely of a corporeal character. It is a body, spherical, perfectly homogeneous and motionless. And all we say about birds, about snakes, about men, all that would refer to deceptive appearances. Change would be an illusion. The real is one and motionless. That is traditional Parmenideanism. Aristotle remarks, at this point, that "their view will be, not that all things are one, but that they are nothing." [3] The world has disappeared. Meyerson fully agrees. The history of the sciences is here to show that we do want to explain the world and that to explain is to identify, and so, at the limit, if we had fully succeeded in our endeavor to explain the world through identification, we would have nothing left except a homogeneous unit as anticipated by Parmenides; the reality of the world would have disappeared. You can now understand the purpose of this remarkable title, *Identity and Reality:* conflict between what the scientific mind wants to do and what is imposed upon us by our experience of reality. Meyerson never resolved the conflict. In his later writings he said that identification as a method of explanation is obviously something unsatisfactory since to explain by identifying is ultimately to explain away, to suppress that which is to be explained. He wrote with a bit of melancholy and more firmness in his very last months that the human mind has not yet found any better way of explaining than the process of identification. I hope we shall perhaps see some day that the process of identification may occur in a variety of ways. I think it is pretty obvious both from the history of the scientists—see Meyerson—and from our reflection on what we do everyday in daily life in our endeavor to overcome our bewilderments, that identification takes place mostly through the material cause, the bearer, the subject (in Greek "subject" is *hypokeimenon,* that which lies under). Are you baffled by novelty? Do not worry— that which lies under it is one and the same. We explain by bringing forth an identity, which identity is in a distinguished and privileged fashion that of the underlying, the subject, the bearer—and that is what material cause means.

For the materalist, then, change is more apparent than real. Underneath there is a big unchanging ball, the Parmenidean sphere, or there are little tiny balls, the atoms of Leucippus and Democritus. Their atoms are simply a pulverization of the sphere of Parmenides—the metaphor is not in Aristotle, I am helping a little. That one big sphere is a little too hard to swallow. It works a lot better if instead of having one big homogeneous sphere, you have indefinitely many tiny invisible ones. But the principle is the same.[4] You have a subject that does not really change and you arrange it in indefinitely many ways. That is the everlasting foundation of atomism, which started at the latest at the time of Plato—I say "at the latest" because I suspect that in India it already existed sometime before the century of Plato—and today it is going on and it can go on and on for ever. We can question the reality of such particles but whether they are real or symbolic, we shall keep using them. For, when we want to explain phenomena, the use of a permanent substrate presents obvious advantages.

In conclusion, I wish to raise the problem of the kinds of change. Here we had better start with Aristotle who has the most embracing view. Soon we shall see how other thinkers restricted the nature of change. In Aristotle there is a distinction between what is properly translated as mutation and what is more properly translated as motion. A bit of warning concerning a sometimes very vexing problem of terminology. When people like Aristotle need extreme precision in vocabulary they make the necessary distinctions but on the next page where they no longer need such precision, they use synonymously the two words whose meanings were sharply distinguished by the page before. Thus, in Aristotle, the word for "motion," *kinēsis*, and the word for "mutation," *metabolē*, will cover both. The same thing happens to a number of technical words. For instance in Book Three of the *Physics* when Aristotle comes to the subject of chance he has two words which I would translate as chance and fortune. It is very unfortunate that translators have not agreed on the way to translate those technical terms. That causes endless difficulty. These two words have definitely distinct meanings. A fortune event, for Aristotle, is an accident which imitates the object of a rational intention. I go to the market place to buy food and here I catch the fellow who owes me money and that I have been unable to find for a number of years. If I had known he was at the market place I would have gone there even if I had nothing to purchase. It is a fact I went there to effect a purchase and I caught my debtor. It is a fortune event, an accident which imitates the object of a rational intention. A chance event, for Aristotle, is an accident which imitates the object not of a rational but of an infrarational intention. For instance, if a lioness having lost her young jumps across a creek because she is pursued by hunters and on the other side of the creek finds again the trace of her cubs, that is a chance event in the sense of Aristotle: an accident which imitates an infra-rational finality. That distinction is very sharp; turn the page, and the word that I translate as chance will stand for either of the two, or the word translatable as fortune will stand for either of the two. That is what Aristotle does and all his commentators do the same also. This may be inevitable. So, we must not overdo the distinction of mutation and motion, sometimes it is observed and sometimes it is not. We shall observe it here. Mutation in the precise sense designates generation *(genesis)* and destruction *(phthora)*. Some would say generation and corruption. Others would say production and destruction. I suppose they are the simplest and safest expressions. Motion *(kinēsis)* comprises 1) local motion, that is, change from place to place *(phora)*, 2) qualitative motion whose proper name is alteration *(alloiōsis)*, and 3) quantitative motion which may be either increase *(auxēsis)* or decrease *(meiōsis)*. Remark two things. First, in either form of mutation the change takes place between contradictories, from non-being to being or from being to non-being, e.g. in death we go from being a man to not being a man. This is positive reality and the negation thereof. Thus, it goes from the negation of a positive reality to that positive reality in generation and from a positive reality to the negation thereof in destruction. For, to be and not to be are opposed as contradictories. Second, in *kinēsis*, motion properly so called, the change is from one positive reality to another positive reality. That is, it takes place between contraries. One positive reality is not the contradictory of another positive reality. They are not opposed as to be and not to be but as two incompatible ways of being. All this is simple and crucial. Thus, in local motion, to be here is one positive reality, to be in another room is another positive reality, and no matter how much we may hate it there is incompatibility between these two positive realities. Local motion is change from one positive reality, being in a place, to another positive reality, being in another place. The same goes for a qualitative motion such as alteration. To be pale is to have a positive complexion and to be pink is to have another positive complexion. When the anger of a man is suddenly aroused there is change from pale to pink, that is, an *alloiōsis,* an alteration in the sense of Aristotle. For qualitative change the examples will either almost always or always be taken from the realm of the living. I think it will always be so. In regards to quantitative change, consider first the growth of a plant. It has a certain height and this is a positive reality. When it reaches another height, which is also another positive reality, it has undergone an increase. If a man weighing 180 pounds shrinks to 150, he has gone from one positive determination to another positive determination. This is a decrease, a quantitative change. A little later we shall inquire into the subject of these changes. You may have remarked that in the case of the three kinds of motion the subject has the character of determination, of completeness, of fullness and of independence. It is one and the same man who goes from one room to another room. He is a man independently of his being in a room. The same for quality and for increase and decrease. On the other hand, in generation and in destruction, where is the subject? What is the subject? Next time you go to the countryside and stop in the front yard of a farm, observe the chickens and the pigs wallowing in their food troughs. Are not those animals factories that produce eggs and bacon out of grain? Where is the subject of that change? That is what Aristotle called "first matter," the great indeterminate, the thing that has no nature of its own. But this is an anticipation.

As I have remarked, Aristotle uses a single term to refer to three kinds of change: local, quantitative and qualitative. This term is *kinēsis*, motion. (Let us call it motion$_1$). He uses another word, *metabolē*, mutation, to refer to generation and destruction. But, on the other hand, when he needs an all-embracing word to designate every type of change he has nothing else than the word "motion." (Let us call this motion$_2$). We have here not only a question of terminology but an interesting problem of logic, which I will merely state without discussing. "Motion" may mean just motion, or it may include both motion$_1$ and mutation. In the former case, we have only a question of words; in the latter, a question of logic: is this division a univocal or analogous one? In other words, is local motion a motion in the same sense of mutation? Obviously not. We have two diverse meanings that are however connected. Compare this with the contrary situation of the word "bat" which means both a flying mammal and a club with which to play baseball; here we have two diverse meanings that are unconnected with each other. We call a term like bat equivocal and one like motion analogous. And so, the division of motion$_2$

into motion$_1$ and mutation is an analogous one. It is tremendously important that we be able to recognize such problems of logic as we run into them in the investigation of nature or of morality or of the world above nature. For, very often our difficulties are due to lack of clarity about the logical issue involved.

One might wonder whether Aristotle considered his division of motion to be exhaustive, i.e., to express all possible forms of motion. If he did, then this division belongs to demonstrative knowledge ("science" in the Aristotelian sense), the logic of which he expounded in the *Posterior Analytics.* Or was he doing what we so often do in our expositions, just picking up the main forms of motions, the best known, the most interesting ones, without being able to show that a fifth sort is impossible? In this case the division would not be scientific but dialectical, dialectics being the kind of probable research which goes on in dialogues among cultured people who know the respectable opinions and who know enough formal logic to get the best out of these respectable opinions. I think Aristotle believes it to be exhaustive and scientific. And I think that his commentators do the same. They consider every change to be one either of place or of quality or of quantity or of substantial determination. What makes me think so is that there are innumerable Aristotelian texts, perhaps not so much in Aristotle himself as in his commentators, to show that there is no motion that pertains directly to the category of relation. Consider the case of a man who last year had no children but now is a father: he has become a father. It looks as if he underwent a change from not having a certain relation to having it. The motion however does not take place in the relation but in another category, and as a result of such a change there arises a new relation. But this does not place motion in the category of relation. Such a conclusion is, I think, well grounded in the texts of Aristotle.[5] Has he ever given a satisfactory demonstration of the exhaustiveness of this fourfold division? That I could not tell you. But that it is exhaustive in his mind, of that I do not think there is any doubt.

NOTES TO CHAPTER TWO

1. *Metaphysics,* A, 982b 12.
2. The law or principle of parsimony is the name given to various formulations of a methodological rule, of which the best known statement is attributed to William of Ockham: Entities should not be multiplied beyond necessity. The idea behind it is that since nature always achieves its effects in the most simple and direct fashion possible, the most adequate theory or explanation of phenomena will be the one that uses the fewest postulates.
3. *Physics,* 185b 23.
4. *Physics,* 184b 15–22. See also *On Generation and Corruption,* 325a 4–34.
5. *Physics,* 226a 24.

Readings (through Chapter Five)

* *Aristotle,* edited and translated by P. Wheelwright (New York: Odyssey Press, 1951).
* W.D. Ross, *Aristotle* (New York: Meridan Books, 1959).
  V.E. Smith, *The General Science of Nature* (Milwaukee: Bruce, 1958).
  Thomas Aquinas, *Commentary on Aristotle's Physics* (New Haven: Yale University Press, 1963).
* A.G. Van Melsen, *The Philosophy of Nature* (Pittsburgh: Duquesne University Press, 1953).

Chapter III

THE SCIENCE AND PHILOSOPHY OF INERTIA

Throughout the history of ideas about nature one cannot fail to be impressed by the fact that local motion is usually taken as a distinguished case among the four types of motion. Local motion is granted a special position by most philosophers. Now, as we have seen, the great contrast in the interpretation of nature is between Aristotelianism and mechanism. Aristotle is the thinker who interprets nature in terms of natures and the mechanists interpret nature in terms of space. But both recognize that local motion enjoys a privilege, that it occupies a distinguished position, that it is not on the same footing as the other kinds of motion.

Let us see first how the question appears in mechanism. The best we can do is to take Descartes, who is very outspoken on this subject. There is a celebrated passage in his unfinished treatise *On the World,* which was composed in part prior to the *Discourse on Method.* It is a lovely diatribe against anything Aristotelian on the subject of motion. You have to keep in mind here that Descartes did not know much history of philosophy and tended to lump together in one mass all the philosophers prior to his own malformation. Thus he once said that Aristotle held to the same views as Plato did and that quarrels between Aristotelians and Platonists are affairs of persons and of schools rather than of philosophies.[1] It is hard to imagine how at a time when the writings of both Plato and Aristotle were easily available such ignorance could be professed so candidly, but

Descartes almost always is very honest and means what he says. It shows however that he had never read Aristotle or Plato. And why had he not? Obviously because he was thoroughly disgusted by the kind of teaching he was given in his adolescence. Then too, if you want a perfect exemplification of his general statement that he does not accept in his physics any principle which is not accepted also in mathematics, you have it here. I translate freely and literally: "They [the philosophers] confess that the nature of motion is very little known to them and in order to make it in some way intelligible they have not yet been able to do better than to use the following terms: *motus est actus entis in potentia ut in potentia est*. To me these words are so obscure that I am compelled to leave them here in their language because I could not interpret them." This Latin is an exact translation of Aristotle's definition of motion: the act of a being in potency inasmuch as it is in potency. Descartes declares it completely devoid of meaning. Had he known German he might have used the word so often employed by the Vienna Circle: *sinnlos*, meaningless. He continued: "But, on the contrary, the nature of the motion of which I want to speak here is so easy to know that the geometers who of all men are those who tried the hardest to conceive very distinctly the things that they have been considering, held its nature to be more simple and more intelligible than that of their surfaces and their lines, as clearly appears from the fact that they explain 'line' by the motion of a point and 'surface' by the motion of a line." Now what is Descartes doing with this beautiful quotation from Aristotle? Aristotle in this celebrated definition of motion taken from his *Physics*[2] is speaking about nature. Descartes is competing with him and boldly substituting for a (to him) completely unintelligible notion of motion one which is used by geometers and is most clear. It is clearer even than the concepts of surface and of line. It is so much clearer than those very clear concepts that geometricians define a line by the motion of a point and surface by the motion of a line. That is what Descartes has in mind when he thinks of motion.[3] Is it local motion? Of some sort, no doubt. Though there is nothing substantial and certainly nothing qualitative about it, it is a motion in place. But in what sort of place? Motion of what in what sort of container? But now, as you can see, we are no longer in physics for there is a substitution of mathematical for physical entities. At the very time when he is endeavoring to interpret the physical world, Descartes applies very faithfully his principle not to accept any principle that would not also be accepted in mathematics. The motion of which he is going to speak in his mechanistic theory of the physical world is a motion mathematically understood. Indeed, it is so mathematical that it has a character of primariness in the genesis of mathematical concepts, being the motion involved in the definition of such elementary, fundamental and primitive mathematical concepts as those of surface and line. Thus, he goes on, "The philosophers posit several motions which according to them can take place without any body changing place." In Aristotle of course a motionless thing can change from one qualitative disposition to another qualitative disposition. It would not be in motion from here to there but it would change from one disposition to another. "But I know of no other motion than the one which is conceived of more easily than the lines of the geometers: the motion according to which bodies move from one place to another and occupy successively all the spaces that are between these two places." [4]

If we want to understand the geometrical space in reference to which Descartes, and more generally, mechanists, are defining motion, I suppose that the most fundamental feature to bring forth and on which to concentrate is homogeneity. Take this word in its literal meaning. Homogeneous means "of the same kind." Take for instance this room that I happen to be in and my motion from one part of it to another, and consider my motion in a concrete physical way. Suppose I move close to the window. It will mean that there is a change of temperature, also a change in lighting, also a change in relation to the other people in the room. Moving from here where I am to the window would be to move from what is qualitatively of one sort to what is qualitatively different. There is no homogeneity between the physical place where I am sitting now and the place where I would be sitting if I move to the window and sat with my back in that nice welcome sunshine. Nothing of the kind in the space of the geometer, which is the great creation of the Greek mathematicians but reportedly of Archimedes in particular. It is the conception of a container whose parts are all alike, whose parts are homogeneous. So we may raise the question: to move from one part of this container to another part of the same, is that to change? To move from a place which is of a certain kind to another place which is of the same kind through places that are of the same kind, is that still to change? We recognize here the idea of Meyerson, in this everlasting effort of the human mind to cut down to a minimum, to reduce if possible to an illusion, the reality of change. One very effective way to do it is to reduce change to motion in homogeneous space, that homogeneous space which can be more properly called, I think, Archimedean than Euclidean, and which was rediscovered with great power of intuition and treated as a physical reality by the founders of the new science whose names are Galileo and Descartes.[5] I think that the core of this problem is whether to move in a homogeneous space is to change. What we can say at once is that if that is all the change that takes place in nature, it is superficial. When I consider my picture of only twenty-five years ago and look at my face in the mirror today, I notice a change which is a little deeper than what happens to me when I move from this seat, for instance, to that seat, although there may be even complete homogeneity here. So, if change boiled down to moving from place to place in an absolutely homogeneous space, what could be said at once is that it would be a pretty superficial process. But there is something deeper than that. How are the parts of this homogeneous space going to be defined and distinguished from each other? We speak of moving from one part of space to another place. How is one place defined and distinguished from another place in that homogeneous space? Distinguishing one place from another is very simple when there is no homogeneity. For instance, suppose we are in the desert five miles east of a certain spring or in that particular small part of the desert which is green and where plants, cattle and men can get the most valuable thing in the desert, water. Here the meaning of place and position is quite clear. For there is qualitative diversity, functional diversity, teleological diversity. If you want to treat your rheumatism keep away from the spring; if you are thirsty move from the driest part of the desert to the least dry. Clear. But where there is homogeneity, what is it that will define a place and distinguish one place from another place? It will be a system of relations. Mechanistic space involves, I think quite constantly in the history of thought, a relational definition of place. Aristotle's definition of place is not purely relational. We shall see later what it is. In mechanism, and in particular in the mechanism of Descartes, it is unmistakably so. To be in a place is to be the subject of a system of relations such as being on the right-hand side of, being on the left-hand side of, being above or being below. A place is defined by a system of relations and by nothing else.

In Aristotle, place cannot be defined in a purely relational way, and this is a weakness of Aristotle's *Physics*. It is a weakness, perhaps, but in the mind of Aristotle that is what keeps it physical. With Descartes, we are no longer in physics, of course. We are in a geometrical interpretation of the bodily world and behind it we have a philosophy which, I think, has to be idealistic for this mathematical space to constitute the physical universe; with such a space we have to reinterpret reality in an idealistic sense. Descartes does not mean to do that. He is not aware of the idealistic dynamite with which his philosophy is loaded. He is by intention a realist. There are quite a few things which he says are only in our minds, like colors and sounds for instance. They are what Locke was going to call secondary qualities. For Descartes those things have no existence except in our minds. But this is a real world and if rational life disappeared from the face of the world, these various things about us will keep moving and keep going on independently of the fact that there is no longer any mind to contemplate them. That is what Descartes means. Now, what is the intention of Cartesianism? Mark the difference between the intention of a philosopher and the intentions of his system. They may coincide. In that case we say that the philosopher knows very well where he is going. And they may not. A philosopher whose intentions coincide fully with those of his system will be an extraordinarily successful and conscious system-builder, but that certainly does not happen often. Generally, there is some amount of discrepancy between what the philosopher really wants to do and what he accomplishes and delivers to his posterity. Descartes delivers to posterity a philosophy which cannot help developing into radical idealism for several reasons. One of them is this: How would it be possible to reduce the physical world to a system of geometrical patterns without one day declaring that matter does not exist except in a mind? I remember what Gilson said one day, "You may start with Descartes if you please but then you have to end with Berkeley or with Kant." Whoever begins as an idealist will end as an idealist; it is impossible to compromise with idealism. If you yield the principle the conclusion also will be idealistic. As I am speaking of that to you, I am seeing things more clearly than I did in the past. If only a month ago you had quizzed me on the idealism of Descartes I would have told you that the crucial point is that in Descartes things are not known directly. What is known directly is some mode of the psyche, and then from the characteristics of the mode of the psyche we infer what the characteristics of the real thing necessarily are. Now, as I am speaking to you I realize better than I ever did that this mathematization of physical nature, this reduction of all motions to a local motion which is itself pure transition from one part of a homogeneous space to another part of the same homogeneous space is impossible without ultimately assuming that the physical world is made of mind stuff. That would be very obvious if only we were a little clearer about the meaning of mathematical abstraction. There is in mathematical abstraction a contribution of the mind which is such that if the whole of nature is interpreted mathematically without residue—I say "without residue," for there are many things in nature that everybody interprets mathematically but grants that there is a residue—then I do not see very well how we could escape a final disappearance of nature into mind. Such a disappearance is very radical in Berkeley and not so radical in Kant.

Now we are ready to see what happens philosophically when the principle of inertia is promulgated by Galileo and by Descartes. We must know the historical background to understand what the question is. The historical background is that for an Aristotelian a thing at rest does not call for any particular explanation or cause. No particular causal reference is needed to see why a rock is sitting here. But if the rock goes up in the air and falls down, for instance, on your skull, you have a right to ask: Who in the world has been throwing that rock up dangerously and perhaps with a vicious intention? That is rather definitely Aristotelian. It is perhaps a cheap way to expound Aristotelianism. If we went very deeply into the theory of causality in Aristotle, especially if we went beyond physics, that is, into metaphysics, then the problem of rest would perhaps be a little less simple. I am describing the prevalent view of the old physics at the time of Descartes and Galileo, a view which was certainly not without a foundation, even if it was somewhat over-simplified. In other words, the problem of causality in Aristotle is primarily relative to motion. I do not say "exclusively." If Aristotle has a metaphysics it should not be. It is today fashionable, especially in the school of Professor Gilson whom we were quoting a few minutes ago, to hold that Aristotle has only a philosophy of nature and has no metaphysics. That is slightly exaggerated. He certainly has a metaphysics. What is true is that he has been stung by the example of Plato, and perhaps also by the audacity of his Platonic youth, and in his mature phase at the time when he composed the treatises that we know, he is an extremely cautious metaphysician who does not want to say more than he really knows. In the *Phaedo* Plato talks very loudly even when he is not sure. From all we know of the early works of Aristotle, as a young man he did the same, but in his mature age he became extremely cautious. That is not the same as to say that he has no metaphysics. He certainly has a metaphysics but, no doubt, it is incipient, not completely shaped by any means; he is aware of it and that is how he wants it to be. In Aristotle metaphysics or first philosophy comes after physics. That is one absolutely fundamental characteristic of any Aristotelian metaphysics. It comes after a physics. In Descartes it comes before. Read the *Metaphysical Meditations* and you will see that those meditations have to take place before physics begins to be worked out. They are necessary for the grounding of physics, though this grounding can be done rapidly and then forgotten. You know that the foundation is firm and you can go ahead with the construction of what is really interesting—physics. That is Cartesian. In Plato, metaphysics is certainly possible without physics—but I should not speak that way. I ought rather to say that Plato calls first philosophy dialectics for very good reasons; in a way it corresponds to what will be called metaphysics in Aristotle, but it is not a thing that comes after physics and is achieved through the construction of physics. I think it is perfectly safe to say, as Duhem did many times, that there is no physics in Plato, that a physics in Plato is impossible, that the only sciences are mathematics and dialectic and that as a member of the school of Plato it was the greatest of all audacities on the part of Aristotle to have inserted in the system a third science, namely, the science of the mutable, of the changing, of that which is considered deceptive in Plato: the physical world.

To sum up: If we leave metaphysics out of the picture for a second, it is correct that for Aristotle the problem of causality is first of all the problem of tracing motion to its origin. A causal problem does not appear as long as things are at rest. One would manifest itself from the metaphysical standpoint; I do not think it does from a physical standpoint. So, let us say, and this is very safe, the causal problem in Aristotle is relative to motion, not to rest.

Here is the great revolution. Give a certain amount of motion to a ball on a plain surface and after a while it will no longer move. This can be interpreted in two ways: the old and the new. The old is that the amount of motion imparted by the cause has become exhausted after a while and so because the cause is no longer there to originate motion, there is no more motion—that is the old philosophy, the one of the pre-Galilean and pre-Cartesian era. The other explanation is friction. A ball that is set in motion is rubbing itself against a smooth surface which is not infinitely smooth, and is also running into the resistance of the air—that is not exactly as if you threw it into clay or into butter; the resistance of clay or butter would stop a ball much faster, in a much shorter time, but the resistance of the air is not negligible. And if you combine the resistance of the air and friction against a surface which is far from infinitely smooth, you have a perfect explanation for the fact that motion does not last indefinitely. Now, make the surface smoother, give the ball the same push and you will see the difference. How much farther it goes! And, again make it smoother and you will see that as friction is less significant the ball goes farther and farther with the same initial impulse. Now, suppose we reach the limit, with no friction and no resistance on the part of the environment: the thing in motion stays in motion just as the thing at rest stays at rest. That is the principle of inertia. You know that this principle was a big success. Does it express physical reality? Or is it a construct which works both theoretically and industrially? Let us leave that question out of the picture. For Galileo and for Descartes, as we have seen, it was not a construct. It was an expression of absolute truth. Remember how according to Duhem a physical theory is nothing else than a construct which accounts for appearances. For Duhem and for a number of physicists of his time, the Ptolemaic geocentric system and the Copernican heliocentric system are two hypotheses which save the phenomena with an unequal amount of smartness, economy and efficiency. If you ask Duhem or Poincaré whether the sun really and truly moves around the earth or the earth around the sun, they will consider you a Philistine and one who does not know what physics is all about. They overdid it badly and it really became the fad of the time. But one thing that Duhem has very nicely shown is that, whereas in Plato, in Symplicius, in Maimonides, in Thomas Aquinas and in various writers of the sixteenth century astronomical theories are considered hypotheses whose duty is to save the appearances, on the contrary, Galileo and Descartes want to be taken literally. Descartes especially would have been greatly agitated had he ever come to suspect that any part of his system had only the significance of a hypothesis, a construct capable of saving the appearances but not expressive of the absolute truth about nature. Whatever may be the real status of the law of inertia, it is considered by both Galileo and Descartes as expressive of the absolute reality and truth of nature. Motion is a state as well as rest. Motion is no longer change. So, we are back to the ideas of Meyerson: the law of inertia is a big step along the line of explanation through identification. Explanation through denial of change: what a formidable step! Change has been reduced to local motion, which moreover has been deprived of its physical character! Descartes told you that, when he said that he knew of no motion except the one which is so geometrical as to be more primitive in geometry than line or surface. And at the same time that strictly geometrical view of motion is held to be expressive of the absolute reality of physical things and this motion is no longer a change from one state to another state, it is a state just as much as rest.

You learned the law of inertia when you were school children. That is the progress of the human mind: things which could be conceived only by a genius in the early seventeenth century have become very familiar to children in the twentieth century. It does not mean that children of this century are smarter; it means that there is a communication of concepts, a communication of tools which after all is something quite valuable. But how are we going to interpret inertia? Is it a construct which works both in the systematization of facts and in our endeavor to control nature? That is one way to say it, especially with those stories of satellites. It is very handy to imagine that once a missile is no longer subjected to the resistance of the air, no matter how light, then nothing interferes with its prevailing in its state of motion. One day it will be borne out that that space you read so much about these days, is not purely geometrical: there are a few particles that move around in that space so that after a certain number of generations satellites may get worn out. But, you see how tempting it is to reach for the limit and to imagine complete freedom from friction and the uninhibited assertion of this law of inertia: a body in motion remains in motion as naturally as it would remain at rest if it were at rest and undisturbed, or better yet, the undisturbed body in motion remains in motion as naturally as the undisturbed body at rest remains in rest. No cause is needed to keep it in motion anymore than a cause is needed to keep a resting body at rest.

At this point it might be both interesting and illuminating to digress a bit in order to consider the evolution of Maritain's ideas on this subject. In one of his early works, *An Introduction to Philosophy*, which was already on the market when I was a student, he suggested that the Galilean principle of inertia could not express the truth about nature.[6] He thought it could be only an hypothesis in the sense of Simplicius, because he saw in it a contradiction of the principle of causality: a motion without a cause. That was the first phase, during the youth of Maritain.

The second phase comes with the publication of *The Degrees of Knowledge* in 1932. A footnote reads:

The motion of projectiles that caused so much difficulty for the ancients, could perhaps be explained by the fact that at the first instant of motion, and by reason of that motion, the qualitative state which, existing in the agent, is the immediate cause of motion, passes into the mobile thing *secundum esse intentionale* (speaking in an ontological key; we are deliberately using terms that do not belong to the vocabulary of mechanics). With this point of view it would become possible to hold Galileo's principle of inertia to be valid not only from the point of view of physico-mathematical science (at least, according to Einstein's mechanics, for an ideally supposed space absolutely without curvature), but also from the point of view of the philosophy of nature.[7]

I am not at all sure what to make of this, but note how he uses the conditional in all those sentences. Further, it is a footnote which is the place where you put suggestions of which you are not too sure. Besides too, there is a reference to an intentional way of existing in nature. Well, I have worked on that subject also. I have been very close to suggestions of that kind and I do not think anybody has gotten very far. Intentional existence in the case of cognition, that is something quite clear. It is a very late expression needed to convey views stated by Aristotle with utmost clarity in a number of passages, especially in the third book of the treatise *On The Soul,* where he keeps repeating that the soul is in a way all things. So, things have two ways of existing, one in nature and one in the soul. In the Middle Ages the expression "intentional," which is very felicitous indeed, was coined to express that quite distinguished way according to which things exist for the second time, when, having existed for the first time in nature, they come to also exist in the soul. Where does the expression come from? For many many years I have suspected the translators of Avicenna of having originated that application of the word intention, which is primarily relative to the will: good intentions, bad intentions—to cognition. I have never been able to ascertain it completely, but that will soon be done because I have old students who have plenty of genius, who know Greek, Arab and whatnot, and who are working on the subject. Now, if you study the very texts of Aristotle you see that if existing intentionally means something in the case of cognition, it must also mean something related in the case of nature. There is no doubt that Aristotle is not an Epicurean. He sometimes looks like one and his translators and historians sometimes did not mark the difference between an Aristotelian form and an Epicurean simulacrum.[8] But the images of the Epicureans are physical things, miniatures, just smaller than the big things, whereas for Aristotle sensation results from a thing impressing on the senses what we may call its ideas, which are not things but non-substantial entities of a certain sort, namely, the things accidental forms. So, if we use intentionality to designate Aristotelian ideas, I think we have to speak of intentionality not only in cognition but also in nature. Clearly, this is a very obscure subject. It has been further obscured by a lot of myths which are due to nonmicroscopic investigations down to the eighteenth not to say nineteenth century. I have written a few things on that subject.[9] The historical situation is extremely confused. For people of Antiquity or of the Middle Ages or of the Renaissance just as well, what is very small is supposed not to be physical. In Descartes, for instance, light knows no time. In our language we would say that the speed of light is infinite. But Descartes had absolutely no way to ascertain the finiteness of the speed of light. That is typical of every thing prior to eighteenth and even of nineteenth century natural investigation. What is very small is supposed not to exist. You can see in medieval texts about sensation—I have read a few of them—that because a physical process cannot be detected (except with our very refined instruments) they hold it inexistent and speak of an intentional process. Thus, the intentional is burdened with a number of tasks that we know to be performed by ordinary physical processes of things which are just a little smaller, so small that they cannot be observed without modern and ultra-modern instruments.

There is something else to consider; there is the theory of instrumental causality. In instrumental causality there is a special way of existing, and that way of existing sometimes has been called intentional. If you take a painting by Leonardo da Vinci you cannot question that everything in it has been caused by the brush. Absolutely nothing has jumped on the canvas without the instrumentality of the brush. The same for the hand of the master—that is also an instrument. That is really embarrassing, yet we have to grant that everything in that masterpiece—beauty, spirituality, splendor, metaphysical sense— all that has been caused by the brush. But really, is it? Certainly: in the order of instrumentality. In a transitional and fluent way, all that existed in the brush. You cannot question that. He did nothing without his brush. Now, that is perhaps what Maritain has in mind here. About the same time as he, on a lower scale, I was working on the same topics, and I am not under the impression that we have gotten very far. I do not think he has worked anymore on those subjects since then. I did, but I am not too proud of the results. I still cherish the hope that one day the study of extra-sensory perception and such strange phenomena might be a way to the exploration of that nonconventional way of existing, which you find in the sensorial ideas of Aristotle and which you find in instrumental causality. Of course, that can be done only when the extra-sensory perception men get serious, and right now they joke about it so much that we had better refrain for a while.

About six or seven years ago I heard Maritain speak again on these subjects. By that time his mind having matured—you know that philosophers mature very very slowly, the age of fifty is about the beginning (a consolation for philosophers over 50!)—his mind having matured and broadened, he had come to wonder if it is absolutely impossible that motion be a state and that the Galilean law of inertia should be more than a construct, an expression of a physical possibility. It is always interesting to see how the state of such a difficult issue can evolve in a great mind. Thus, the first stage was: "It cannot be anything else than a construct because it would be contrary to the principle of causality." The second stage: "Perhaps it is not so simple, perhaps a special way of existing is involved here." And the third stage is a reconsideration of those old views about rest and motion, not being so sure that persevering in motion should be more of a causal problem than persevering in rest. Here, I am mostly raising problems, for these are difficult subjects, though worth considering.

We have noted something very striking: local motion, which in mechanism enjoys a privilege obviously connected with the essentials of mechanistic philosophy, also enjoys a privilege in Aristotle whose philosophy of nature, it can reasonably be said, is the opposite of mechanism. That coincidence is very striking. Let us try to see why local motion enjoys such a privilege in Aristotle. We have already remarked that there are two changes, generation and destruction, which are not motions except in a broad sense of the term. Generation goes from not-to-be to to-be and destruction from to-be to not-to-be. Motion properly so called, in Aristotle, is between two positive realities. Not between a positive reality and its privation but between two positive realities, between contraries. This is the theory of contrariety in nature, an obscure point in the philosophy of Aristotle.[10] What I can do at this point is just to mention that the notion of contrariety, in Aristotle, has two unequal meanings. In one sense, two contraries are always two positive realities which are incompatible in one and the same subject at the same time, like, for instance, to be sitting in this room and to be sitting in another room. The two are positive and they are incompatible, we specify, in one subject at the same time. A man can be sitting in this room and another man in the next room. Or, a man could be sitting in this room now and then walk to the other room and sit there. In neither of these cases do we have contraries. That is the broad notion of contrariety in Aristotle: positive realities that are not, to use a rare word, compossible, possible together. There is another notion of contrariety in Aristotle which is very problematic, which is related to the qualitative aspect of Aristotelian physics, and which is precisely that which was most ardently rejected by the new science. When he speaks with the greatest propriety, Aristotle calls contraries two positive realities which are extremely remote from each other within the same genus. That definition is a fairly literal translation of Aristotle. For instance, if you take colors, are blue and red contraries? In the broad sense, yes. One and the same surface cannot be blue and red at the same time. That is clear. Now, in a more intense sense, the black and the white are contraries. Within the genus, color, the greatest distance is not between the blue and the red but between the black and the white. Here, contraries have to be within one and the same genus. You would not say, for instance, that the black is contrary to the acute. We would not say that a color is contrary to a sound. In regard to colors, sounds and such qualitative genera Aristotle has a leading notion, in his interpretation of nature, that distances are more or less great.

We find the greatest distance between the black and the white or the cold and the hot, and these are contraries in the proper sense. And in the becoming of the physical world he attributes an absolutely decisive importance to the contraries. But you can see at once that that theory of the action of the contraries in physical becoming is one of the parts of Aristotelianism that is going to be badly hit by our common notions of the relativity of such qualities as the hot and the cold.

To avoid one unnecessary possibility of misunderstanding we may note that when Aristotle says "within the same genus" in his definition of the contraries, he means the last genus, the one under which there are ultimate species, under which species there are only individuals. We have here a very simple question of logic. Here is a genus (e.g., quality) comprising logical wholes (e.g., color, sound) which are species if you refer them to the genus, and genera if you refer them to ulterior species. Now as we subdivide, a time comes when we have only species and those species are called ultimate because they contain no other species but only individuals. In other words, a last genus is one that contains species which are not themselves genera. Aristotle refers to this when he speaks of contrariety because if he referred to a genus containing other genera, then that obviously would not work. The broadest genus would be quality and in quality we have not only sounds and colors which cannot be said to be contrary to each other, but we also have sciences and arts and shapes and all sorts of things. Thus, there is little doubt that the genus of which he speaks in the definition of the contraries is the last genus.

In our endeavor to show why, in Aristotle also, local motion enjoys a privilege, we first remark that such changes as generation and destruction which take place between nonbeing and being or between being and nonbeing are not motions properly so called. A motion properly so called takes place between two positive realities. Now, we still have three of them: local motion, alteration and quantitative motion (growth and decrease). Why the privilege of local motion in the nonmathematical interpretation of nature in Aristotle? I think that the ground of this privilege of local motion is continuity. We are going to see that motion demands continuity between these two extremes, the two positive realities between which motion takes place. Between these two there must be indefinitely many steps—I am expressing myself almost metaphorically because we do not mean indefinitely many jumps from one point to another point but precisely the opposite of that, a noninterrupted development. Is there continuity in growth? In decrease? In alteration? Aristotle obviously thinks that unqualified continuity is the privilege of local motion. As a result, it is more of a motion, more fully a motion, and he considers also that it is the *sine qua non* condition of every other motion. The other forms of motion would not be possible if they were not borne along or supported by a local motion. I am not completely clear about all of that. But there is one thing which is completely clear in my mind: the continuity of local motion is clearly connected with the homogeneity of quantity.

To understand why that is so we must consider how we define quantity. Of course, quantity is one of those primitive notions which do not admit of a definition properly so-called. But when a definition properly so-called is not possible, a substitute for a definition may still be very helpful. In the case of quantity there is a helpful substitute for a definition of quantity. I consider a thing, for instance, this table. I decide, first, to consider it qualitatively, that is, in terms of color. Then what strikes me if I consider this table qualitatively is that when I move from one part of it to another part of it, I move through a heterogeneous field. For instance, here, I can see an area characterized by the reflection of the windows. No such thing here. Lighter spots of color—darker spots of color. Here, moving from part to part means moving to what is other. In this way, we see how quality is a principle of differentiation, of specification, of heterogeneity. Now, suppose I consider this table not as a bearer of qualities but just as a surface. This is a quantitative consideration, no doubt. I abstract from those spots of color, I even abstract from the shapes—that would be qualitative also—I consider it just a plane surface. I move from one part of it to another. Clearly, I move from the same to the same. We are as close as anybody can be to a definition of quantity. It is really, indeed, something intriguing and very puzzling. In quantity you have a distinction of parts that are alike, homogeneous, of the same genus. Take the table as nothing else than a surface. Moving from one part of it to another part of it means moving from what is of one genus to what is of the same genus and yet those parts are distinct, they are distinct by being outside each other. Here is the miracle of quantity. It effects distinction by mere extraposition. Now, this is not a definition, but it is a very helpful substitute for a definition. You can see that it is not a definition when I say "outside each other." You could quiz me on the meaning of outside and I suppose we would be caught in a perfect circle. We have to include "outside" in a definition of quantity, but if we wanted to explain the notion of outsidedness we would have to include in it the consideration of quantity. That is what I mean when I say that this is not and cannot be a definition properly so called, a definition according to the logical rules. This is a substitute for it which is valuable. It is valuable because when I say those parts are distinct by being outside each other, the word "outside" is an appeal to a sensory and imaginative intuition, the intuition of outsideness that suffices in itself—and there can be nothing else when we define supreme genera such as quantity.

To sum up, looking for the reasons why, in Aristotle as in mechanism, local motion enjoys a privilege, I find in him this answer: motion must be continuous and it is only in local motion that there is unqualified continuity. I have tried to make one more step by connecting the continuity of local motion of which Aristotle speaks with the homogeneity which really defines quantity.

You see how in diverse philosophical systems we are using the same words to speak of deeply different realities. When we read Descartes and speak of quantity, it is a good thing that Descartes makes it unmistakable what he means. It is quantity mathematically considered, quantity in the state of mathematical abstraction, but that Descartes will never come right out and tell you. He will never tell you that because the basic postulate of his physics is that the physical world is sufficiently explained in mathematical terms, which makes sense only if there is no difference whatsoever between 1) quantity mathematically understood, quantity as treated mathematically, and 2) quantity which is a property of real physical things. In Aristotle, the difference is very clear and very significant. Here quantity is a property of real physical things. It is this real physical quantity which, so far as I know, through its property of homogeneity, makes it possible for local motion to achieve complete continuity. But why should motion be continuous in order to be completely faithful to its notion? That is a further point you will eventually want to look into.

NOTES TO CHAPTER THREE

1. "That road is to seek out the first causes and the true principles from which reasons may be deduced from all that which we are capable of knowing; and it is those who have made this their special work who have been called philosophers. At the same time I do not know that up to the present day there have been any in whose case this plan has succeeded. The first and principal whose writings we possess, are Plato and Aristotle, between whom the only difference that exists is that the former, following the steps of his master Socrates, ingenuously confessed that he had never yet been able to discover anything for certain, and was content to set down the things that seemed to him to be probable, for this end adopting principles whereby he tried to account for other things. Aristotle, on the other hand, had less candour, and although he had been Plato's disciple for twenty years, and possessed no other principles than his master's, he entirely changed the method of stating them, and proposed them as true and certain although there was no appearance of his having ever held them to be such. But these two men had great minds and much wisdom acquired by the four methods mentioned before, and this gave them great authority, so that those who succeeded them were more bent on following their opinions than in forming better ones of their own." *The Philosophical Works of Descartes.* Ed. by Haldane and Ross. (New York: Dover, 1955), Vol. II, p. 206.

2. Aristotle, *Physics,* 201a 10.

3. Compare a remark made by Etienne Gilson in his Gifford Lectures: "It is this fundamental insight which Aristotle is trying to express when he says that movement is the act of a thing in potency in so far as it is in potency. Since the time of Descartes it has become fashionable to scoff at this definition, and certainly Descartes' own seems a good deal clearer; probably, as Leibniz saw, because it altogether fails to define movement. The obscurity does not lie in Aristotle's definition, it lies rather in the thing defined: something, namely, which is in act because it is, but is not pure actuality because it becomes, yet has a potency which tends to actualize itself progressively because it changes. If we look at things instead of words, we cannot fail to see that the presence of movement in a

being reveals a certain lack of actuality." *The Spirit of Mediaeval Philosophy* (New York: Scribner's, 1940), p. 67.

4. *Oeuvres de Descartes,* ed. by C. Adam and P. Tannery (Paris: Cerf, 1897–1957), Vol. II, pp. 39–40.

5. I recommend especially on this subject the studies of A. Koyré, *Etudes Galiléenes* (Paris: Hermann, 1939) and *From the Closed World to the Infinite Universe* (Baltimore: Johns Hopkins, 1957). A very good historian of the sciences, he understood particularly well the meaning of Galileo and Descartes as founders of the new science. He saw a number of things which escaped the attention of his great forerunner, Duhem.

6. Jacques Maritain, *An Introduction to Philosophy* (New York: Sheed and Ward, 1962), p. 71, note.

7. *The Degrees of Knowledge* (New York: Charles Scribner's Sons, 1959), p. 115. Several years later, Maritain remarked further: "Taking the principle of inertia as established [for the sake of discussion], and even hypothetically granting it a meaning beyond the mere empiriological analysis of phenomena, it suffices, in order to reply to the objection, to note that, applied to movement in space, the axiom 'Everything which moves is moved by another' ought then logically, by the very fact that motion is considered a state, to be understood as meaning 'Every body which undergoes a change *in regard to its state of rest or of motion* changes under the action of another thing.' And thus the axiom remains always true. According to the principle of inertia in its classic form a body once set in motion continues of itself to be moved in a uniform manner or with the same velocity. If then the velocity of its motion increases or diminishes, it will be because of an action exerted on it by another thing." (J. Maritain, *Approaches to God* [New York: Harper and Brothers, 1954], p. 32.)

8. See Cyril Bailey, *The Greek Atomists and Epicurus* (New York: Russell and Russell, 1964), pp. 407–13.

9. See *Introduction à l'ontologie du connaître* (Paris: Desclée De Brouwer, 1934; reprinted by C. Brown, Dubuque: Iowa, 1962), Chapter III. Also see "An Essay on Sensation" in *The Philosophy of Knowledge,* ed. by Houde and Mullally (New York: Lippincott, 1960), pp. 77 ff.

10. On this point see J.P. Anton, *Aristotle's Theory of Contrariety* (New York: Humanities Press, 1957).

CHAPTER IV

THE PHILOSOPHY OF CHANGE

Let us consider first of all Aristotle's celebrated definition of motion given in Book Three of the *Physics*. This is the definition that Descartes declared very sincerely to be completely nonsensical. We have here a fine illustration of the problem of communication among philosophers. Many would say that the definition of motion by Aristotle is one of the peaks of the philosophic genius of mankind. And yet a very great philosopher like Descartes does not understand it and he is sincere when he tells you that he does not. So, one must not be too upset if a renowned teacher says, "I have read the book of so-and-so and to me it means absolutely nothing." It does not necessarily mean that the book is bad or that it has no meaning. It is quite possible that a meaningful statement by one philosopher simply does not reach the mind of another one. The remarks of Descartes about Aristotle's definition of motion are a perfect example of that very sad situation, but it is a situation which is going to last.

Aristotle defines motion as the *act of a thing in potency, considered as such, that is, considered as in potency.* But, what does he mean? We have here a quite interesting epistemological problem. He is defining the most basic of all physical realities. For Aristotle at least, there is nothing closer to the core of physics than the concept of motion. Yet he is defining it in unmistakably metaphysical terms! Note how he uses exclusively terms of an obviously metaphysical character: thing, act, potency, and when he says, "as such," what he is using is the concept of identity. So we can say that he defines movement or motion, the physical reality par excellence, not in physical terms but in metaphysical ones.

I have never read even a brief discussion of this striking fact. This in itself I consider illuminating, because motion is absolutely basic and fundamental in all parts of physics. It is as if the first principles of physics admitted of no physical explanation and if explained at all had to be explained in terms of metaphysics. I think that is the case and I find it perfectly plausible. But I am a little shocked to notice that after having read so much on those subjects I have never seen one page where this remark was made.

The metaphysical terms into which Aristotle resolves motion are act and potency. Because of their analogical multiplicity of meanings these terms lend themselves to all sorts of misinterpretations. But what especially causes difficulty is not act; it is potency. To show what the difficulty is, I know of no better method than to refer to a misinterpretation that I find in Meyerson. I have already expressed my high opinion of Meyerson. I think his works will be classic for a few centuries. It is impossible to show more awareness, clear-sightedness and ingenuity in investigating what science is really about and what the purposes, methods and laws of scientific thought are. As I have indicated earlier, Meyerson contends that science is concerned with explanation and that to explain is to identify. It is this basic principle which accounts for the success of what we may call materialism.We notice changing appearances and we are baffled and annoyed by the experience, but really we need not worry too much since only the surface changes; if we dig a little, we shall find something that does not change. One of Meyerson's illustrations of his basic theory that the scientific mind is constantly looking for a subject or a substrate that does not change is the theory of potency in Aristotle, whom unfortunately he has not understood. That is his weak side. He was reputed to know all the European languages but when he quotes Aristotle he uses the one French translation available at that time which we as students were forbidden to use, as exceedingly inaccurate. We cannot trust him when he speaks of Aristotle but we may use him to show how potency can be misinterpreted.

For Meyerson, when Aristotle says that what is in act right now already existed in potency twenty years ago, he means that what is observable right now is not new, but already was there twenty years ago. In other words, he thinks potency is a force which eventually manifests itself. But that is not what the Aristotelian potency is. Meyerson has simply not understood that point. That is tremendous. If Aristotle's potency were a stable subject that lies under changing superficial events, then, indeed, the philosophy of Aristotle would confirm magnificently the interpretation of explanation as identification, with the danger that ultimately when we have explained everything, we have identified everything and we really have explained *out* everything and we are confronted by nothing, because if ultimately there is no multiplicity and no change, then what have we explained?

What I want to make clear from the beginning is that the Aristotelian state of potency is not a state of hidden existence. When you read Leibniz you will find occasionally the word potency and more often the word virtuality. And it has often been said that the virtuality of Leibniz is an impeded act, a hampered act, an act that something prevents from showing, from unfolding. I think that is what it is in Leibniz, but this is not the potency of Aristotle.

I have developed what seems to me a simple and effective method to explain the notion of potency in Aristotle. Imagine that on one and the same day a baby and a monkey are born. In Aristotelian terms we might say that the baby is a geometrician in potency and that the monkey is not. We mean thereby to express something real in the baby. The difference is real between being a geometrician in potency and not being a geometrician in potency. It pertains to the thing itself, otherwise, a monkey would be a baby. But that potency is real because twenty years later the baby may become a very bright student of mathematics, but I am sure that the monkey will not.

The day they were born we could say, "You see that baby, that is a geometrician in potency." Two decades later however, from a geometrician in potency he has become a geometrician in act. Many other babies born the same day did not, but he did. Does this actualization of potency mean that when the baby was born he already was a geometrician? No, he simply was not one. I give that example because of the punch of this remark that the day they were born the newly born man and the newly born monkey were equally ignorant of geometry. A newly born baby does not know anything about geometry. He knows nothing. So, he does not know any more than the monkey. Nothing is not any more than nothing. There is in them simply an equal ignorance of geometry. You do not find in the baby a hampered, hindered or inhibited actuality of geometry. There is simply no geometry. What is there? A real possibility; a possibility which is not merely logical, but is real; a possibility which exists in one thing and does not exist in others. That possibility really exists in the baby but does not exist in the monkey. Still, if asked "Which one is the better geometrician?" we would answer they are just equally ignorant of geometry.

Meyerson simply did not understand that. He imagined the potency of Aristotle to be a predetermination, a prefiguration, an outline of what is going to show some day. But it is no outline. The newly born baby does not have even a vague knowledge of geometry so far as actual existence is concerned, but by reason of his specific nature he has in him an ability to develop an understanding of geometrical entities. This ability is something real, but it is not actual knowledge. It is just a real potentiality to know, that is all.

When discussing this point one day in a classroom a young lady told me, "How can you know that a newly born baby is a geometrician in potency? He may be feeble-minded." Well, of course he may be feeble-minded, in which case an ability pertaining to the species would be held in check by an individual accident. I do not certify that concretely speaking and considering all the features of individuality this newly born baby is able to become a geometrician. I am considering that which

THE PHILOSOPHY OF CHANGE 63

results from the specific nature. Now, what results from the specific nature may be held in check by an accident pertaining to individual existence either before or after birth.

It would not be so hard to understand what potency means if we were better aware of the logic of this and of many other issues. For one thing, it is clear that terms like act and potency admit of no definition properly so called. Mathematicians have thoroughly popularized the notion that the elementary and fundamental concepts of a science cannot be defined. There can be no regression to infinity in definition any more than in demonstration. Terms like act and potency are too primitive to be defined, but they can still be explained through a number of examples. I am almost quoting literally the famous passage of the *Metaphysics* of Aristotle where he says that indeed those terms cannot be defined but that you can get an idea of what they mean if you consider that potency is related to act the way sleep is related to wakefulness.[1] He also gives a few other examples of that kind.

There is however something else more subtle to consider, a logical problem which really controls the understanding of these issues and of many more or less similar issues. Consider the division of being into being in act and being in potency—this is not a division of a genus into its species, it is the division of an analogical concept into its analogates. To see what this means, we must keep in mind the distinction logicians make between univocal, equivocal and analogical terms. Should a word be used in just one sense, it is called univocal. For instance, if I say that a camel is an animal and that a horse is an animal, animal as predicated of camel and of horse has purely and simply one and the same sense. These are two species of animal, which is thus a generic, univocal term. On the other hand, consider these two sentences: "He left the game because the bat was broken," and "The bat is not a bird but a flying mammal." What about "bat" and "bat"? These are really two words but they are spelled the same and they sound the same. Because of the sameness in the sound and spelling, they can be considered in the material sense to be one word with two diverse and unrelated meanings. It is by accident that we use the same sound and the same letters to designate the bat which is a flying mammal and the bat which is a club with which you play baseball. Those words are called equivocal. Are they also ambiguous? Sometimes. Equivocal is more precise because ambiguous would not necessarily exclude the third sort of term in which we are interested, analogous terms, which convey two or more meanings that are not unrelated. There is the example of "healthy," popularized by Aristotle himself. You say that an organism is healthy, that a diet is healthy, that a climate is healthy, that a sample of urine is healthy and the gait or complexion of a person is healthy. Of course, "healthy," here, conveys several senses. In the case of the organism it means having health, in the case of the climate or the diet it means causing health, though in modern English it has become customary to say healthful rather than healthy in that case, and in the case of temperature charts, urine or complexion, healthy means an effect and sign of health. "Healthy" obviously has three meanings which however are not unrelated. Here is another example: "I found them in a cheerful mood," and "Their apartment is small but cheerful." Can an apartment be full of cheer? Obviously not, but an apartment may be so designed, so lighted and so decorated as to cause a cheerful disposition in those who live in it. Another example would be: "This cabinet is a very conscientious piece of work." Can a cabinet be conscientious? Only in the way urine can be healthy, as an effect and a sign.

There are other forms of analogy. One is the metaphorical analogy which is the best known but not the most important. For instance, we say that an eagle is a bird and that a man of genius is an eagle. Here, there are two senses, related by a transfer. Then there is also the great analogy, that did not acquire a distinct name until very late, the so-called analogy of proportionality which you find in terms like "good." It is perfectly correct to say, "This ice cream is good," "His health is good," "His mother is an awfully good woman," and "God is good." Does the word "good" have the same meaning in each case? Certainly not; its meanings though related are diverse. The importance of being able to recognize these different sorts of analogous terms cannot be overestimated.

One of the great problems of analogical concepts is that of the order among the analogates. Thus, taking "being in act" and "being in potency" as members of a set, we may call this an analogical set.[2] Which of them comes first? Act does, of course. Should we say that the one that comes second is understood through the one that comes first? Obviously. Here is the point: potency is not intelligible all by itself. It is intelligible only through actuality, as a relation to actuality. To say that the newly born monkey is not a geometrician even in potency and that a newly born baby, so far as his specific nature is concerned, is one, is intelligible in relation and only in relation to the actual possession of geometrical science. This great question of the order within analogical sets, how one term of the set is understood through the other one, is a complex question because there are diverse types of analogies and perhaps indeed, indefinitely many types of analogies. Furthermore, the law of orderly understanding does not have the same meaning in all cases. That is why the question is difficult. But difficult or not, we have to tackle it if we want to understand things like the state of potency.

Considering, again, being as divided into "being in act" and "being in potency," note that in the case of "being in act" the common ground, being, is purely and simply asserted. In the second case, in the case of being in potency, the common ground, namely, being, is asserted, indeed, but also negated. That is the secret of analogical thinking. In the case of the second analogate you have an assertion of the common ground coupled with a negation of the common ground. The newly born baby is just as ignorant of geometry as the newly born monkey. Absolutely speaking, purely and simply he is not a geometrician and yet he is a geometrician in potency. You have here a togetherness of the "yes" and the "no," an association of assertion and negation which concerns the common ground. If the common ground were a genus, assertion and negation would concern only the differential factor. Compare within the genus animal, the mosquito, the chimpanzee, the man. I can say that a chimpanzee has differential characteristics that the mosquito does not have. When I understand a chimpanzee I posit something, let us say, the characteristics of this particular species, and simultaneously I deny the characteristics of another species. But notice that the association of assertion and negation concerns the differential. With analogates the association of assertion and negation concerns the common ground. I stop here because I wish only to call your attention to the significance of those logical instruments and to the impropriety of trying to go very far in metaphysics, and more generally, in philosophy, without working out the appropriate logical instruments. When we have to understand a difficult subject like the state of potency, the difficulty pertains principally to logic and if the logical instruments are mastered then it is no longer so terribly difficult.

In several passages near the famous definition of motion as the act of a being in potency as such, Aristotle also declares that motion is indeed an act, but an imperfect one, one that does not have a termination. Let us try to understand this notion of motion as imperfect act. That is a real approach to our main question. We get, here, into another analogical set where the understanding of one member helps to understand the other member. If he calls motion an imperfect act, what would be an example of a perfect act? Let us consult another famous passage of the *Metaphysics:*

At the same time we are living well and have lived well, and are happy and have been happy; if not, the process would have had sometime to cease, as the process of making thin ceases: but, as things are, it does not cease; we are living and have lived. Of these processes, then, we must call the one set movements, and the other actualities. For every movement is incomplete—making thin, learning, walking, building; these are movements, and incomplete at that. For it is not true that at the same time a thing is walking and has walked, or is building and has built, or is coming to be and has come to be, or is being moved and has been moved, but what is being moved is different from what has been moved, and what is moving from what has moved. But it is the same thing that at the same time has seen and is seeing, or is thinking and has thought. The latter sort of process, then, I call an actuality, and the former a movement.[3]

This is typical Aristotelian phrasing of something which is not really terribly difficult. Just think of what happens when being on one side of a hill we are eager to see what is on the other side. We climb up the hill and when we have reached the top of the hill we may sit down and contemplate the landscape and keep contemplating if we please. That example tells the whole story.

Is it possible to be climbing and to have climbed it? Aristotle says that it is not. It is possible, indeed, to come from below to this point and to be still climbing toward a point farther up. What is not possible is to have been climbing to this point and still to be climbing to this point. What is not possible is to have climbed to the top of the hill and still be climbing toward the top of the hill. That is what he means. Likewise, is it possible to have found a truth and to be finding the same truth? If we consider the mind in movement toward, say, a certain conclusion, it is not possible to be in movement toward the conclusion and to have reached the same conclusion; if we have been in movement toward this conclusion, once the conclusion is reached the movement of the mind belongs to the past. And so you can see how many sorts of activities are imperfect acts which cease to be as soon as the term is reached. Among the examples Aristotle gives of this, one is making thin; another is learning. We may have learned one thing and be learning another thing but we may not be learning one thing and have already learned it.

And what are the actualities that admit of the terminal character that motion is denied? That is clear and simple: there are three. One is contemplation. When you find the word theory, *theoria,* in Aristotle, it generally or always means precisely that, to be looking at the result of a search, to be no longer searching but to be considering or gazing at the truth which has been discovered. It should be clear that "contemplating the truth" does not imply that we have reached the end of difficulties, or of mysteries; it does not imply that the whole of the truth is known. If I speak of this it is because so many people when they hear of contemplating the truth believe that somebody claims to have exhausted the possibility of knowing a certain thing or a certain object or a certain field. Let it be understood that when we speak of contemplation, we are aware that we always contemplate and know infinitely less than there is to be contemplated and to be known. But a knowledge that is incomplete in a thousand ways may still be the answer to a certain question. Someone may raise a question a few centuries before the answer is obtained, and one day when the answer to precisely this question is obtained, we do not yet know everything, we do not know the whole about anything, but we have nevertheless obtained a true answer, a true proposition. Grasping a truth which may be inexhaustive in a million respects but is still genuine is all that is needed for the notion of contemplation as terminal act, perfect act, act of a subject in act, to make sense. So, contemplation is the first and the best known case.

A second case is joy. Joy is normally the act of an agent in possession of what he is looking for. Contemplation and joy are two things which go very well together. Then too certain forms of love have the character of terminal acts. You can see very well that it is possible to love an object in its absence or while striving and struggling toward it. That is possible; it happens all the time. But when the loved object has become present, when there is achievement of what was striven for, when there is actual company of a beloved person, love does not cease. You all know the speech of Socrates, in the *Symposium*, where love is described as essentially restless. It reads *as if* (with Plato, of course, you never know for sure what he means) love expired as soon as it is satisfied. It reads very much as if love were necessarily and essentially an imperfect act which died out as soon as the term is reached. Is that precisely what Plato means? Does he exclude the possibility of love in rest, in peace, in contemplation and joy? I shall never dare to say that Plato excludes anything, but it is certainly the plain meaning of the theory of love sketched in the Symposium. But we ought never forget that these are dialogues so that you never know how definitely and definitively the writer adheres to the theory which is proposed.

Charles Sanders Pierce has also written on this subject of thought and motion. He discusses it in the famous paper entitled "How to Make Our Ideas Clear" which is supposed to be the foundation of American pragmatism and which appeared in 1878 when Pierce was around forty, just beginning to mature. In it he says:

> . . . the action of thought is excited by the irritation of doubt, and ceases when relief is attained; so that the production of belief is the sole function of thought.[4]
>
> Thought in action has for its only possible motive the attainment of thought at rest. . . .
>
> But, since belief is a rule for action, the application of which involves further doubt and further thought, at the same that it is a stopping-place, it is also a new starting-place for thought. That is why I have permitted myself to call it thought at rest, although thought is essentially an action.[5]

Now, we must understand his terminology, his language. He cautions the reader that the terms "doubt" and "belief" are too strong for his purpose since these words generally refer to religious or other grave discussions. By "thought" he obviously means something like what is designated by "learning" in Ross' translation of Aristotle. It is thought in movement, thought searching, thought moving from premises to conclusions or from particular instances to a general proposition. Thought is action that is "excited by the irritation of doubt" and when belief is attained it dies out. Although he speaks of thought in action and of thought at rest, he sees very well that what he calls thought in action is for the sake of thought at rest. That is quite clear. But he denies that thought at rest is a kind of action and that is a nice way to make the difference between a pragmatic view and an Aristotelian view. Identifying thought in action with thought in motion, Pierce sees very well that thought in motion is for the sake of thought at rest, but the thing he misses is that there is such a thing as a restful activity, such a thing as an activity by way of rest. That is absolutely clear and most important in Aristotle, the notion of an activity by way of rest. His examples as we have seen are contemplation, joy and the loftiest and the happiest forms of love. Keep these things in mind. They are not so easy to find in the reading of Aristotle because of the uncertainty of the translators. One will call thought at rest actuality. Another uses the word activity. A third prefers the word act. Yet another one would call it achievement or accomplishment. The most foolish will use a word literally derived from the Greek, such as entelechy for instance, that nobody understands. If it were possible to establish conventions for the translations of those key words the difficulty of philosophic studies would be greatly eased, but I am afraid it is impossible.

To illustrate my point further let us take one man whose vocation is to create something and another whose vocation is to know—let us say a painter and a philosopher. Let us suppose they get started the same day, January first, one putting spots of color on the canvas and the other inquiring into a problem. Let us suppose too that they reach the term of their endeavors on November first. What is the difference? For the painter, *qua* painter, *qua* artist, all is over because painting is motion, painting is action by way of motion. And when the painting is finished, the action which produced it belongs to the past. Now what will the painter do? He can take a vacation, he can sit down and admire his painting, he can sell it or he can start painting another canvas. It is clear that as painter of this particular painting he is done; his action is over; his action no longer exists, but belongs to the past, because it can exist only by way of motion; it can endure only by way of motion. Clear. The philosopher also started working on a problem on January first and the first of November he has found an answer— although generally it is more like one generation or a few centuries than from the first of January to the first of November— but let us assume he has established the basic though perhaps not exhaustive answer to his question by then. On that day thought in motion is over. It belongs to the past. Research belongs to the past. And Pierce would say that thought in action belongs to the past. But is it not clear that the contemplation of the actually possessed truth is also thought in action, though not thought in motion? It is thought in action by way of rest. For the painter, all is over. But we would say that, for the philosopher, all begins. Everything before was preparatory. Life really begins when it is possible to think not by way of motion but by way of motionless contemplation. Is that an action, motionless contemplation? Read Aristotle. His key theme is that it is an action indeed, that it is not necessary for action to be by way of motion, that it is possible for motionless action to be terminal, finished, perfect.

With this we can understand better the motion which is described as the act of an imperfect subject, as an imperfect act. I repeat the famous definition: act of a being in potency as such, that is, considered precisely as in potency. After the preliminaries we just went through, this definition that Descartes confessed so sincerely his inability to understand should be rather clear if we bring in just one further consideration, that in this imperfect act which is motion, in this act of a thing in potency precisely *qua* in potency, you have to consider two relations of potentiality. Let us first look at a case in which there is only one relation of potentiality. Take for instance an artist working on a piece of marble and trying to bring a certain shape out of that noble material. If he is interrupted in his work we have something unfinished, incomplete and not in motion. What is the difference between a thing in motion and a thing arrested in its development, in its progress toward a certain term? That is all important. It involves whether we are going to catch or miss the reality of motion.

There is a great twentieth century philosopher who contended that traditional philosophies, or most of them, missed the reality of motion and caught only inappropriate substitutes for it. He died in 1941 and his name is connected with the contention that the misfortunes of philosophy originated in improper methods as a result of which instead of grasping motion, we only grasp as a substitute for it a succession of motionless instances. I am of course referring to Bergson. He first presented such a criticism in 1889 in his doctoral dissertation, the *Essay on the Immediate Data on Consciousness,* published in English under the title *Time and Free Will.* His subject is a psychological one, his main interest being motion in consciousness. The leading idea of the book is that the objects of consciousness are not really perceived as they are because the fluent reality of consciousness is arrested by destructive analysis. His great American friend, William James, welcomed those views with elation. James, who was twenty years older than Bergson and a charming man, had the modesty to declare to the world that he was a disciple of Bergson. The world fame of James was such that Bergson was very quickly promoted because of James' adherence to what seemed to be most basic in his ideas. James calls motion as considered in the field of psychology the stream of consciousness, an expression which gained considerable currency for a couple of generations but has now died out. This is unfortunate, since it is beautifully expressive of what James and Bergson were pointing out. Later, at the beginning of the century, in the book entitled *Creative Evolution,* Bergson pursued his search for the genuineness of motion and endeavored to go beyond the motionless substitutes that an inappropriate analysis produces. He applied his ideas to physical nature and so worked out a philosophy of universal mobility which has often been likened to that of Heraclitus at the dawn of Greek thought.

To know whether the extreme expressions of universal mobility that you find in *Creative Evolution* belong to the deep thought of Bergson, you have to read his final work, *The Two Sources of Morality and Religion,* published in 1932. It is very clear that this book has been written by the author of *Creative Evolution.* I remember very well the day when I was told, "Bergson's book of ethics is out." Bergson had not published any important book since 1907. It was known that he was working on a book of ethics and it was feared that he might very well die without having completed it. I was about to give a lecture and to catch a train. I gave the lecture and before I jumped into a streetcar to catch my train, I first ran into a bookshop to buy the book and I was already flipping the leaves as I was entering the streetcar. What struck me first was that this book was obviously written by a disciple of the man who had written *Creative Evolution*. That was the first impression, but perhaps not the most profound. Whether Bergsonism finally remains a philosophy of universal mobility is not so sure: perhaps it does not. It strikes me that some aspects of Bergsonism as the philosophy of universal mobility are corrected in this last book of his.

At any rate, Bergson helps us to see the problem. An imperfect act may be nothing else than the situation of a thing arrested in its development, motion or movement. Then, what is it that distinguishes motion from arrestation? It is a second relation of potentiality. Let us return to our example of an artist making a marble statue. When the shaping is interrupted and arrested, there is a certain potentiality of the marble that has been actualized. Suppose it is a statue of a human body. Suppose the two arms are shaped and the legs are not. Well, there is in act a certain possibility of that material, but further possibilities simply have not been actualized. (I take this example from the domain of human art for the sake of simplicity. It is obvious that those notions have no particular connection with human art.) So, we have behind us, behind the present situation, an actualized potency. The act of this potency may be more or less complete. There is act of a certain potency and that is all. What is the difference if the process of shaping is not interrupted and arrested, if we consider the shaping itself precisely as it goes on, precisely as distinct from arrested motion? The difference is that there is another relation of potentiality. Some potency has been saturated but more potency remains, and the saturation of some potency results in the possibility of the saturation of more potency. We have a relation of potentiality behind and a relation of potentiality ahead of us, and the two are one. Or, let us say rather, the two exist together and are one because the one ahead is unceasingly being transformed into the one behind. This is very precisely the reality of motion as distinct from all the substitutes that, as Bergson showed I think very correctly, a number of analytical methods considered instead of motion itself.

Let us take another example: an arrow in motion toward a target, the arrow of Zeno. The problem is to see what it means to say that motion is the act of a thing in potency as in potency. To put it in another way, what does final actuality mean in the case of an arrow in motion toward a target? It is the reaching of the target, obviously. That is what is terminal; that is what the term is here. In fact, words like term and target are often very closely related. Suppose the arrow drops on this side of the target and no longer moves. Since it is only half of the way, the actuality of the arrow as a missile aimed at the target is incomplete. Half of the way has been done, so some potency has been saturated and all we have is arrestation, arrested motion. We no longer have motion. Now, consider the arrow in motion, not yet reaching the target but with its motion uninterrupted, unimpaired. Part of the way is covered and because of that we can speak of act and actualization. We can say with propriety that a certain potency has been actualized but the act which is the actualization of this potency is itself potency to further actualization. That is what "the act of a thing in potency as such" signifies: these two relations of potentiality, which exist together in the thing in motion. What distinguishes the thing in motion from the thing stopped is that the actualization of the potency behind is one with the ability to stand further actualization. I must admit that I am not particularly proud of the way I am expressing myself, although I think what I am saying is correct.

We started out by bringing forth the contrast between the philosophy of nature and the science of space. We pointed out a fundamental issue: Should the physical world be interpreted in terms of natures in motion or in terms of space? Of mechanism or Aristotelianism? Many would tell you the failure of Aristotelianism in the sixteenth and the seventeenth centuries has brilliantly demonstrated that a philosophy of nature, an explanation of the world in terms of natures, is impossible. It is by doing analyses like we are doing now that it may be possible to ascertain which one is right, whether mechanism is necessarily more than a method, whether it is true that the only possible interpretation of the physical world is one in terms of space, or whether it is possible to interpret the physical world in terms of natures, and so to work out a philosophy of nature. Just let me call your attention to this, that we would have to change our language considerably if we wanted to be consistent mechanists because our language is constantly postulating that things have natures. In the case of motion, recall the text of Descartes and how unmistakably he asserts that he is not speaking of something different from what the geometricians have in mind when they say that a line is generated by the motion of a point and a surface by the motion of a line. There can be no doubt about it. This is motion already conditioned and processed by mathematical abstraction and one unmistakable effect of mathematical abstraction is to rule out finality. On the other hand, we cannot over-estimate the importance of this very simple remark made in two or three places by Aristotle, that mathematical entities are true and that they can be beautiful, but that they cannot be good. They cannot be good because they are so conditioned by mathematical abstraction that they cannot be good without ceasing to be mathematical and without becoming physical things. Thus, you may desire a triangle, which is then something good, but we are no longer speaking of a mathematical triangle, we are speaking of a gadget that you can buy in an office supply shop and which is handy to draw right angles or forty-five degree angles—that is, a physical triangle; that is the only kind of triangle that is desirable and good. The question of goodness makes no sense once a thing has been treated by mathematical abstraction. Do not be surprised if there is no teleology in mechanism. Mechanism is nonteleological and antiteleological precisely insofar as it remains faithful to the great Cartesian ideal of understanding by motion nothing else than what geometricians do when they say that a line is generated by the motion of a point.

After Descartes declares that under the name of motion he understands nothing else than what the mathematicians have in mind, he works out a notion of motion that obviously has absolutely nothing to do with finality. For, the term of such motion does not have any character of goodness because it is no longer a goal, a target, an aim; in short and most precisely, the term is in no way an end. He is not wrong in being aware that this concept of motion is sharply at variance with that of Aristotle who sees motion as directed toward a state of actuality which is a state of perfection and also an end. In regard to this point you can easily get all information you need by reading Aristotle's *Metaphysics* (Bk. 9, Ch. 8). You find there a discussion about the priority and posteriority of act and potency. Which one comes first? It depends on the point of view. If you speak of becoming, one is a baby before one is a mature man; so, in becoming, at least in the becoming of individual things, the state of potency comes before the state of actuality. Now, what about the point of view of perfection? Aristotle subdivides perfection into form and end, and declares that obviously in both ways act comes before potency. Act is more of a form than potency is, and act is more of an end than potency. He shows that the end character of act is particularly clear in those acts, like seeing and contemplating, which last when their term has been attained.

I think we have now about what we need to understand the great crisis at the time of the Renaissance. Try to apply teleological considerations to celestial motions and bodies. The astronomy of Aristotle is a teleological one. He treated the universe like a work of art whose parts are disposed in such and such a way because this is the best possible disposition. Are we really able to know why it is good that the moon should be rotating about the earth? You see how in some cases at least, the teleological interpretation of local motion seems to be impossible. We can see the extreme arbitrariness of the teleological considerations of pre-Galilean, pre-Newtonian, pre-mechanistic astronomy. In most cases questions of finality are unanswerable when local motion is involved. So, if we want to discuss the finality of motion, what do we do? We take examples which are *not* from the order of local motion. We consider rather the growth of a plant, for instance, which is a quantitative and also qualitative development. There, we sometimes can perceive very clearly telic relations. Why is it good that some chemicals are heavier than the air and so descend, while others are lighter than air and so go up? Aristotle believes he has answered those questions. It may be one of the main reasons for the failure of Aristotelian physics when confronted by the mechanistic interpretation of the world.

In the light of this, it is of some interest to note that in *Identity and Reality* Meyerson refers to the notion of teleology as an irrational one. However, to see what he means by this we must be clear about his terminology. For this, we must keep in mind his fundamental theme. Contrary to Comte, he holds that the sciences do endeavor to explain reality. To explain however is always to identify, to show that diversity is more apparent than real and that change is a superficial event which really vanishes on closer examination. As a result, he says, science is constantly producing mechanistic theories, that is, theories that reduce the world to space, apparent diversity to arrangements of space and apparent changes to re-arrangement of the parts of space. If that job could be done thoroughly, then the universe would become rational, would be entirely consonant to our reason, would be explained. But it would also be explained away. Hence the title of the book, which calls attention to the conflict between identity and reality.

What then does Meyerson call irrational? In several parts of his work, he shows that the great project of explaining the world mechanistically is held in check. There are several cases in which the real resists that endeavor to rationalize it. As he describes this rationality however, it is a combination of rationalism and of mechanism.[6] Such a notion he perhaps got from Descartes and from Spinoza, but it is also very much in the spirit of nineteenth century science. It is, certainly, not an indisputable interpretation of rationality and explanation. When then he speaks of irrationality, we must not forget that the negation contained in irrationality refers to precisely this conception of rationality. Thus, he finds an irrationality for instance in irreversible processes. One of the great discoveries of the nineteenth century is that you can take heat from a hotter source and pour it into bodies that are at a lower temperature, but it is impossible to reverse the process. You cannot take heat from bodies at a lower temperature and carry it to a body at a higher temperature. This is one way to phrase the principle of Carnot, the second law of thermodynamics. It has an historical and philosophical significance in the development of physics because, so far as I understand, according to the postulates of classic mechanics, the system of mechanics which is roughly common to Descartes and Newton, if it is possible to take a material point from a certain location to another location it should be equally possible to take it back from the other location to the first one. That is what is not possible in irreversible phenomena, the best known of them being the important fact expressed by the second law of thermodynamics. Imagine a ship at sea. How nice it would be to get heat from the surrounding ocean so as to move the engine of the ship: the ship could go ahead and leave behind iced water. But that is what is impossible; that would be reversing an irreversible process. Now, when you speak of teleology as an irrational it is usually in connection with living beings. Meyerson remarks [7] that up to now it simply has not been possible to reduce the living processes to mechanistic arrangements and re-arrangements. There is something there which resists explanation through identification. Then, when sensation appears we have a third irrational, and I suppose that we could count a few more.

These irrationals are what maintain the reality of the world, without them there would be a universal identification after the fashion of Parmenides. But when everything has disappeared into that Parmenidean unity, we know that we are no longer speculating about nature in Aristotle's sense. We are rather speculating about a non-nature, the nothingness that is left after nature has disappeared.

When Meyerson speaks of teleology as irrational, he does not then at all mean what those words taken out of context would seem to indicate. To understand what he means, you have to keep in mind his notion of rational and of reason. On this point, however, he definitely is ambiguous. Is his "reason" to be understood in an idealistic sense, in the sense of Kant? Well, the whole critique of Kant is designed to explain that which is not empirical in science. And the direction of that critique is rather clear. I think there is no over-simplification in saying that in Kant the scientific object is rational, universal, necessary. The scientific universe is orderly because the scientific object is shaped by the law of the human mind. That is a typical expression of Kantian idealism, though not of idealism in the sense of Berkeley, which is what Kant was criticizing. The Kantian doctrine is not an absolute idealism because the human mind contributes only the shape or form of knowledge. There is a content which comes from a reality independent of the human mind. No doubt about it, Kant is not an absolute idealist. The meaning of rationality in Kant is rather clear; it is a contribution of the mind which shapes data into objects. There would never be scientific objects without that shaping of empirical data by the constitutive forms of the mind. When Meyerson speaks of the exigencies or requirements of our reason, when he says our reason demands this and that, does he mean it in a Kantian idealistic sense? I do not know.

I would ask the same question concerning Michael Polanyi. He had been teaching at the University of Manchester for a number of years and used to be a very successful chemist, but for the last twenty years or so has devoted his mind and his excellent style to the human problems of science. In a rather important work, a book entitled *Personal Knowledge,* which came out in 1958, he seems to do to quite an extent what Meyerson does. Remarkably, he does not seem to have read Meyerson. He is arguing against the popular interpretation of positive science as something statistical, impersonal, communicable, objective, authoritative, final and so on. Throughout the book in diverse ways he brings forth in answer to scores of questions the proposition that real science is not just recording what has been observed nor is it anything like what popular positivism believes, but it involves rather a sort of instinctive taste: affinities for certain arrangements, inclinations towards certain conceptions, a certain sense of duty, a commitment of the person to the sciences, things like that. What is at the bottom of this philosophy? Is it something like those rational forms of Kant which shape the scientific object? That is not clear at all. The inspiration is certainly not the same as in Kant. And, yet, in a number of places, you wonder if in a different inspiration, in a widely different language, it is not something like the *a priori* forms of knowledge in the sense of Kant which reappear, as they seem to have done also in Meyerson.

In the light of all this, I would like to specify one conclusion that I would draw. According to Aristotle, if local motion is taken, not in a Cartesian manner, but concretely and physically, finality is an unavoidable and necessary aspect of it. Factually however, are either Aristotle or we able to definitely establish the goodness of local motion? For instance, can we without arbitrariness express that for which it is good that the moon should circle around the earth? Some have thought that possible. But I would suggest that we cannot, and that there may be many forms, ways and phases of local motion which simply escape teleological interpretation by the philosophy of nature. I would not however deny the feasibility of every such attempt. Take this very clumsy view of Aristotle that light things go up, and heavy ones down. For him it is very clear that it is good that it should be so—indeed, if there were not something like a natural motion of heavy things towards what we call down, is it not obvious that the world would scatter and that the organized forms of physical existence would be impossible? So, I suspect that there are very general cases in which the finality of local motion does not escape us completely. Not so long ago, I would have said bluntly that the teleology of local motion escapes us and that, consequently, local motion cannot be treated from the point of view of physical philosophy, that it has to be surrendered to the mathematical interpretation of nature which does not care for teleological considerations.

And yet I just have to look around to see that there is something good about the heavy things adhering to the surface of the earth, so that there may be some very general aspects of the teleology of motion which do not escape us completely. That is as far as I would go along that line. On the other hand, however, I would deny that teleological considerations are reserved for philosophers and that teleology is absent in scientific facts.[8]

We have teleology in departments of knowledge which are certainly not philosophic, e.g., in biology. You know that when a biologist makes a speech or writes a preface, he curses teleology, tells you that finality is a primitive notion, that is, pre-logical, infra-rational and shameful. Then, as soon as he goes to work, he describes an organ and asks what it is good for: this is done constantly in biology and I think that it is done apart from philosophic consideration.

We have seen that Aristotle, when he wants to be particular, uses the word *"kinēsis,"* customarily rendered by motion, to designate the three cases in which there is change from one positive reality to another positive reality and the subject of the change is a being in act. It is a thing actually existing which leaves a place to get to another place, loses a quantitative determination to get another quantitative determination, or drops one quality to gain another. What about this concept of motion as divided into local motion, quantitative motion and alteration? Does it have a unity of univocity or one of analogy? Is motion as predicated of local, quantitative and qualitative motion asserted in one and the same sense or does it convey a diversity of meanings? If one says that it is univocal because the second and third are reducible to the first, that would be quite logical. The problem then would be whether they are so reducible. As we have pointed out, this is one of the main aspects of the clash of physical theories in the early seventeenth century. For Galileo and still more definitely for Descartes, every motion is reducible to local motion, which they interpret in terms which presuppose a mathematical treatment. Similarly, although mechanists in general may distinguish these sorts of motion, for them motion is univocal for the good reason that the second and the third have no distinct nature. They are superficial processes which on closer examination vanish into the first.

This division of motion is an Aristotelian theory, so it is only natural to ask what Aristotle thinks of it. For him, they are not reducible. On the other hand, I am not absolutely sure concerning his view on the unity of meaning of motion as divided into these three. He has not to the best of my knowledge expressed himself on this subject. I say, "to the best of my knowledge," because you know how daring it would be to assert that such a proposition is not implied somewhere in the not very voluminous but immensely compact work of Aristotle. What I can say is that if the univocity of motion is connected with the reducibility of the second and third to the first, then, for Aristotle, it should not be univocal, because they are not so reducible. I am however more interested in raising this textual problem than solving it.

The general criterion which distinguishes a unity of analogy from univocity is that when you move from one term to the other, there is in the first an assertion of a common ground which receives a qualified negation in the other term. That is the most precise criterion I have been able to work out in all my career, although I have always been interested in the problem of analogy. This is what I would mean and the reason why I think that, for Aristotle, the division is indeed an analogical one. He repeats a number of times, as we have said, that local motion enjoys a primacy because motion to be itself—you see that we are pointing here to the common ground, not to the differential factor—must be continuous. Now he finds continuity in an unqualified sense only in the case of local motion. But in regard to living things, he does not pay much attention to the case of increase and decrease; instead, he compares local motion and alteration—that seems to be the relevant comparison for him. In alteration, however, he does not find that continuity which he considers essential to motion. So, he asserts a feature pertaining to the common ground, continuity, in an unqualified way of local motion and in the case of alteration he combines its assertion with a negation, inasmuch as for him it is owing to a background of local motion that there is some sort of continuity in alteration also. He does not totally reject continuity in alteration, otherwise it would not be motion at all, but he does combine it with a negation. However, that is the criterion of analogy. So I rather think that this division is an analogical one for him. You can see that such matters are often not completely clear. To decide whether a certain division is univocal or analogical is always very important. But what is very important is sometimes also very difficult. Here, we shall simply point out that the analogy in question is one of proper proportionality. It is an analogy not of the type healthy and healthy, cheerful and cheerful or eagle and eagle, but an analogy of the type good and good and good.

Let us broaden our concept and include mutation (metabolē, i.e., generation and destruction), in which change is not from positive reality to positive reality but from nonbeing to being or from being to nonbeing. Here, the thing which changes is not a being in act. It is merely being in potency, matter not in a relative but in an absolute sense, "prime matter." It is a pure "out of which" that has no nature of its own and owes whatever nature it has not to itself but to what shapes it. With Aristotle, we have a common word for this broader concept. It is motion.

This sort of use of the same word in various senses you can expect at all times in Aristotle and perhaps in all philosophers. Bergson, having been accused of using the word intuition in several senses, one day explained with a modesty that was not only conspicuous but real, that he would never liken himself to any of the great philosophers but that Spinoza also had used the word in at least four or five senses. When you find kinēsis or motion in Aristotle you have to ask this question: Is it a case in which he is interested in specificity or not? The meaning of kinēsis will vary accordingly. That is how Aristotle speaks. That is how all philosophers speak to some extent. Absolute specificity of language in philosophy is something like the ideal of algebra and perhaps too of symbolic logic. However, experts are quite agreed that it would just kill the philosophic mind, so we have to put up with those diversities of meanings. In the positive sciences, and especially in mathematics, we are concerned with understanding, distinguishing and relating to each other various species of objects. Consequently, univocal concepts are best suited for these areas. Philosophy however, since it is concerned with providing an over-all synthetic view of reality, has to use terms in extended and analogous senses because it is only in this way that it can deal with the relative unity of essentially diverse kinds of things.

Motion, understood in this broad sense, is very certainly an analogical reduction, Now, when Aristotle defines motion as the act of a being in potency as such, the context makes it unmistakable that he includes mutation. So, this celebrated definition of motion is the definition of a highly analogical concept. We already have quite a bit of analogy data. But that is not yet all. Take an activity such as contemplation, which is not a change but just the lasting existence of an accomplishment. Throughout world philosophic literature you will see it designated occasionally as a sort of motion. This usage is in Aristotle. It perhaps began with him and it goes on and will always go on. So, we have a third analogical understanding. The problem then is to see according to what kind of analogy contemplation, an activity not by way of motion, can be called a motion. Let us first see if it could be one of proper proportionality. In this sort of analogy all the subjects of which the analogical term is predicated are that term intrinsically, properly. A good moral action is good intrinsically and properly. A good mother is good intrinsically and properly. And so is good ice cream. Now, we do not have motion intrinsically and properly in contemplation, so it is not called motion by an analogy of proper proportionality. It is without doubt a metaphorical analogy, one like eagle and eagle. If I say of a man of genius that he is an eagle, I know that he is not an eagle. In terms of reality he simply is *not* an eagle. And a pure activity, an activity not by way of motion but by way of lasting accomplishment, such an activity purely and simply is *not* a motion. It is simply called such by a metaphorical analogy. What this means is that there is an order in analogical sets and that the name of the better known is sometimes transferred to the less known. The first kind of activity I know is very certainly activity by way of motion. Activity by way of motion is so certainly the first kind of activity to be known that many people have a very hard time to realize that there can be activity in rest. We have seen an expression of that in Charles Sanders Pierce. That is how we speak; being more familiar with activity by way of motion there is a transfer, (in Greek, *metaphora*), of the name designating the more familiar case of activity, whose name is motion, to the case of the less familiar form of activity, say, contemplation, joy or the happier forms of love.

In this extended, analogical sense motion will not necessarily involve potency. In a terminal act, in an act not by way of change, like contemplation, there is ordinarily a subject which is actuated, which is perfected by a certain act, and in this restful activity you still have a relation of subject to perfection which is a relation of potency to act. You always have this relation of potency to act in contemplation, except in the case of God. That is what Aristotle said very clearly in his famous dictum concerning the divine thinking, that it "is a thinking on thinking," [9] since God is an act of thought whose object is not distinct from itself. Here, there is pure act and no bearer. The thing which is God is strictly identical with its act of intellection. It is an everlastingly existent intellectual flash. Here, contemplation no longer involves or admits of any potency, but where there is not this identification in the contemplating subject of a subject and his act, there exists a relation of potency to act. So, what distinguishes activity not by way of motion is not precisely the absence of potency but the absence of actualization, of process, of transition from a more potential state to a less potential state. That is the difference.

As long then as we remain in the domain of change or transition of any kind, we can speak of motion, no doubt analogically but with propriety. However, the day I call an act of contemplation or an act of joy or an act of possessive love motion or movement—which is indeed very often done, not so much in philosophy as in mystical literature—then I speak by metaphor. There is a non-metaphorical analogy here but it is not that of motion. It is the analogy of activity. If I spoke of activity then I would speak with propriety though still analogically. If I call contemplation or possessive love or joy, motion or movement—and that is done very much when technical precision is unneeded or even harmful—here the analogy is metaphorical and must be taken for what it is. However, if contemplation and emotion are considered not as motions but as activities we can say that we have an analogy of proper proportionality. To put it another way, the division of activity into activity by way of motion and activity by way of motionless actualization, is an analogy of proper proportionality.

In sum, it is clear that the broad concept of motion has a unity that is neither univocal nor simply verbal but analogous.

NOTES TO CHAPTER FOUR

1. *Metaphysics,* 1048a 36.
2. See Y. Simon, "On Order in Analogical Sets," *The New Scholasticism,* XXXIV (1960), 1–42.
3. Aristotle, *Metaphysics,* ed. by W.D. Ross (New York: Random House, 1941), 1048b 25–34.
4. *Collected Papers of Charles Sanders Pierce* (Cambridge, Mass.: Harvard University Press, 1931–58), Vol. V, pp. 252–3.
5. *Ibid.,* p. 255.
6. E. Meyerson, *Identity and Reality* (New York: Macmillan, 1930), especially Ch. 9.
7. *Ibid.,* p. 63. See also pp. 311–19.
8. For a more detailed consideration of teleology in the cosmos, see Henry Van Laer, *Philosophico-Scientific Problems* (Pittsburgh: Duquesne University Press., 1953), pp. 133–46. See also J. Owens, "Teleology of Nature in Aristotle," *The Monist,* Vol. 52 (1968), 159–73.
9. *Metaphysics,* 1074b 34.

CHAPTER V

THE REAL AND THE
IDEAL IN NATURE

One of the most important aspects we have to consider regarding motion is the problem of its reality. That motion is real is not at all doubtful, although it has been questioned by the Eleatics and we have seen that all mechanistic philosophies tend to reduce the reality of motion to a minimum. However, although the Parmenidean view that motion is only an illusion may have very deep roots in the nature of our reason, as Meyerson held, empirically it remains the most untenable of all propositions about the physical world. This is paradoxical. We can state that the reality of motion is not dubious and yet there is in our notion of motion something unreal. There are some difficult problems here.

Consider the paradoxes of the Eleatics, for instance. Think of their famous questions, such as: "Where is the arrow that is flying toward a target?" In the light of the definition of motion such as we analyzed it, you cannot help but come to the conclusion that there is in our concept of motion a unity that is not supplied by the real itself. Just think of the arrow between the bow and the target. We say that it is in motion between the bow and the target. We can reduce the space traversed again and again, but unless you want to stop the arrow, to represent it as not in motion, you have to contribute a link or connection or tie between that which no longer is and that which is not yet. This is not very difficult to see. What is difficult though is to understand how it is possible that the mind should contribute a construction to the interpretation of a thing as real as motion, the most familiar of realities. I think that the precise point to which we must give our attention is this double relation of potentiality of which we have spoken as absolutely characteristic of motion. This double relation of potentiality is what also involves a unity of the before and after, a unity of that which no longer is and of that which is not yet, which is contributed by the mind. This is extremely interesting. In the interpretation of the most familiar of all physical realities, in the interpretation of what is central to all physical knowledge, we find, on the one hand, reality itself, on the other hand, a contribution of the mind which involves us with an obvious danger of misunderstanding everything by not drawing a line between what pertains to the real and what is contributed by the mind.

Of the few examples of such delicate cases, motion is perhaps the best known. Consider a missile M that goes from an initial point I to a terminal point T. You may consider the whole motion from I to T or you may subdivide it into phases. I suppose calculus was discovered when infinity was injected into that subdivision. What is clear is that so long as you have motion you understand the before and the after as together. If you do not, the missile is stopped, arrested. I understand motion only as I hold together the before and the after. They are not together in reality. It is the mind which by contributing a unity of reason renders the fluent reality of motion intelligible. This does not mean that the fluent reality of motion is not something real. It is real. It pertains to the real world. But it is real in its own way and its own way of being real is such that it cannot be grasped in an act of understanding and cannot be expressed in a concept without a unity of the before and the after, a togetherness of the before and the after which has to be contributed by the mind. So, we have a central, substantial, basic and fundamental area of reality that we want to understand, but because of the fluent nature of this reality it is understandable only if it is held in a wide circle of unity contributed by the mind. You might ask why it is so. Why are there some realities that can be expressed only in and by a concept part of which is contributed by the reason? The unfortunate thing is that to the best of my knowledge, the man who will write a paper on the subject is perhaps born, but the man who has written one is not. Apart from Aristotle's discussion of time in the fourth book of the *Physics* and the work of his commentator Averroës, I know practically nothing on this subject, and in particular on the general question of why some realities are such that they cannot be understood except in a concept part of whose intelligibility is contributed by the mind. In the case of motion and in the case of time the ground for this necessity is, I think, clear. It is the fluent or fluid character of the reality under consideration. Where there is fluency, where there is fluidity, intelligibility is at a minimum. I am, here, giving a remote explanation. I do not pretend to make it airtight. It is true but it is remote. It will have to be elaborated on and that would be really difficult. But that is certainly where the need for the contribution of the mind lies, in that imperfect unity of such fluid realities as motion and time.

In Book Four of the *Physics* Aristotle has a famous discussion of the question whether time would exist if the soul did not. In his typical manner he answers, "In a way yes, in a way no." That is how Aristotle proceeds and it is not always easy to make the "yes" and the "no" so clear as to see in what way it is yes and in what way it is no. Time is very closely related to motion and the problem is the same. In order to have any understanding of time we need to tie up together the past and the future, that which no longer exists and that which does not yet exist. This is clearly quite analogous to what we do when we try to understand motion.

There is another example that is much less striking and less immediately important but still very significant. About a half a century ago A.N. Whitehead wrote a lovely little book entitled *Introduction to Mathematics*. Any mathematician can tell you that it is out-of-date. But such a good book, even if it is badly out-of-date, can be read without any waste of time. In it Whitehead says that all sorts of objects can be counted. Numbers can be applied to any sort of thing. You can count men, you can count angels. Whether he believes in angels or not is irrelevant; what is important is that he posits the notion of a purely spiritual entity having nothing to do with corporeal or physical qualities. Whether it exists or not is another subject. Suppose it exists, would it be countable? Suppose that Gabriel, Raphaël and Michael are three purely spiritual entities. I say three—are they three? Can they be added one to another like three chairs, or three men or three marbles? That is highly dubious, to say the least. So, we have to deal here with a case similar to that of motion and of time. For I cannot think of three angels without counting them, but when I consider the nature of spiritual existence and the principles of number, I understand that angels make up a multitude which is not numerable. However, although it is not numerable I cannot understand that multitude without injecting into it, contributing to the concept of that multitude, the form of number. When you count Michael, Gabriel, Raphaël—one, two, three— are you counting noncorporeal entities or are you counting corporeal entities in one to one correspondence with Raphaël, Michael and Gabriel in an imaginary space? That is what I think we are doing. Raphaël, Gabriel and Michael cannot be added to each other. For, additivity is a property of the quantitative. Now, whether we ought to posit or to hold in doubt the existence of spiritual substances is quite a problem. Is it a properly theological issue? Is it only by revelation that the existence of purely spiritual substances is known? Or is there a rational and philosophical way to determine that there exists finite spiritual substances? Is that way demonstrative or only probable? I leave those questions aside. There certainly exists nonquantitative realities like thoughts, for instance. If I consider the judgment of Mr. So-and-so on this subject and that of his neighbor, am I adding to each other judgments or am I not more precisely adding to each other corporeal fictions placed in an imaginary space in one to one correspondence to the judgment of Mr. So-and-so and to that of his neighbor and to that of the neighbor of his neighbor? I think the latter is true. We started with the words of Whitehead on counting men, chairs and angels. We might just as well speak of counting judgments. That would leave out of the picture the problem of the demonstrability of finite spiritual substances.

We see then what causes the need for a mental component. I cannot say that in the case of motion and of time that it is just the fluency of the thing under consideration. It is rather their metaphysical (i.e., beyond the physical) character what makes them reachable only in association, in a one to one correspondence, with something of the physical order. That is a completely different ground for the necessity for a mental component. So, it should be clear that the reality of motion is absolutely in no way jeopardized. In the same way that I cannot understand motion without injecting into the concept a unity that does not belong to the real world, I cannot understand time without injecting into the concept the representation of a unity of the before and the after which plainly does not belong to the real world.

Such problems help us see how such paradoxes as idealistic philosophies are possible. Suppose that a highly philosophically-minded person one day comes to realize that whenever we think of time or of motion, we think of objects which collapse without a contribution of the mind. The temptation will be great to assert that the world of motion and of time is nothing else than our representation. Indeed, our understanding of it collapses and vanishes without a certain something which is certainly contributed by the mind. When those things are analyzed with great accuracy the door is open to such philosophies uncongenial to, let us say, common sense. Nevertheless, it is hard, indeed it is impossible, to live up to radical idealism.

Let us consider the problem of this element or component contributed by the mind to such notions as that of motion and time. I propose to consider it as a particular case of the general problem of the being of reason. This is as good as any other place to try to clear up this general issue which remains so confused in a great many minds. I can think of very respectable philosophical thinkers who have for years been propagating confusion because they were never clear about what a being of reason is.

It is first of all a problem of vocabulary. If you look it up in *Webster's Dictionary*, you will find the expression *ens rationis*, so it is English. There is even a quotation from William James who is a good writer. The plural is *entia rationis*. Although it is English and in Webster's, I must admit that for many years I and my friends who were translating the logic of John of St. Thomas hesitated over using it.[1] Then one day we concluded that if there is a place where the public would not want a Latin-looking word, it is in a book which is a translation from the Latin, so we decided to employ instead "being of reason." The expression is ambiguous; call it bad if you wish, but the Latin is just as bad. "Being of reason" is an expression which might be misunderstood but there is absolutely not the slightest guaranty that *ens rationis* would be less misunderstood. We must get rid of those Latin words which irritate people quite unnecessarily, especially when the Latin does not say anything more than English words do. There are exceptional cases in which there is a very good word in Latin for which there is no exact translation but *ens rationis* is not one. So, to try to rule out misunderstandings we shall not use it.

A being of reason is an object which neither does nor can exist except in the mind in the capacity of object. You have in this definition all you need in order never to do what has been done by so many people: to confuse a being of reason with a psychological reality. That is the ambiguity of the expression "being of reason," but the Latin *ens rationis* is just as bad. Ignoramuses may take it to designate psychological realities, but a psychological reality is an *ens reale*, a real being of a particular kind that is just as real as anything else. Take a man with plenty of happy memories who is unfortunately involved in a head-on collision, so that as a result of brain injuries his memory is gone. A certain facility that he had to remember what he did as a child and as a young man is gone. Something real is gone. We may elaborate indefinitely on the nature of such psychological realities, our sensations, our images, our recollections, our acts of understanding, our acts of reasoning, our concepts and so forth and so on; that these are real things is not questionable. You may say that they are reducible to movements of particles if you are a very staunch materialist after the fashion of a hundred years ago—that is one way to see things. Then psychological realities would ultimately be of the same nature as the so-called physical realities. Real they are anyway, whether you interpret them materialistically or not. A being of reason is that which neither does nor can exist except in the mind and *in the capacity of object*. This is the distinguishing part, the differentiating part of the definition. "In the capacity of object," not in the capacity of disposition, not in the capacity of habit, not in the capacity of memory or image or concept but in the capacity of object.

Let us consider some examples. Beings of reason are found in several domains. There is one where they are overwhelming because they are alone. It is logic. Logical properties are beings of reason. That is the first thing to get in order to define logic and to distinguish it from it unscrupulous neighbors. Logic is surrounded by neighbors that have absolutely no scruples, for instance, the psychology of the intellect, the critique of knowledge and, worst of all, the ethics of thought. These neighbors of logic are always ready to swallow it up. There are on the market indefinitely many books of logic, especially perhaps since the beginning of the nineteenth century, where there is a little logic and much that may be very good in itself but is not logic. However, what is very good in itself and is not logic becomes vicious when it is called logic. What we have to understand here is exceedingly simple. Just take a little fact such as an incident in the jungle. A beast of prey, a lion, devours a deer. That is a real event that does not belong to the logical world; it belongs to the real world. When you have observed a number of the same such facts you are perfectly entitled to generalize and to say that the lion is a carnivorous animal. Here you are no longer considering an individual, real event, but a general property. I would even say an essential one. We approach very clumsily, imperfectly and unclearly such essences as that of lion. If you ask me exactly where this species of a lion begins and exactly where it ends, you know that we do not know those things. Opinions on it change from generation to generation of zoologists. Though we are very uncertain about those things, when I say a lion and a deer, I am sure that I speak of two different things, things that have different natures. Without being able to ascertain their natures with much clarity, when I say "lion" I circumscribe one thing, and when I say "deer" I circumscribe something else. A lion is carnivorous so that if there are too many deer in a jungle it is a good thing to let the lions do their job. And a deer is herbivorous so that if you grow corn it is better to destroy a deer. All that is clear. We are talking about the real world all the time. We start with individual happenings, then we consider, no matter how clumsily, universal types. We speak of the real world all the time. Then a day comes when I consider the proposition: "The lion is carnivorous." That proposition refers to the real world but I may reflect upon the proposition and say, "In the proposition, 'the lion is carnivorous,' 'lion' is subject and 'carnivorous' is predicate." But there are no subjects or predicates in the jungle. Those objects exist in the mind alone. It is as simple as that in principle. The development of those principles may involve tremendous difficulties. In principle it is as simple as that: a lion belongs to the real world, the devouring of a deer by a lion belongs to the real world, the lion's property of being a carnivorous animal belongs to the real world, and when I stop to think that I understand those properties in arrangements of objects my understanding belongs to the real world too. But as I arrange those objects in such a way as to understand them, what happens to those objects in this mental arrangement? They acquire properties that they never have in the jungle or in the desert. We can put it in a slightly different way. The lion and the deer exist twice, in the jungle and as objects in the mind. As a result of the second existence that they enjoy in the mind, they acquire new properties that depend on their first existence but that follow in part too from the distinguishing characteristics of this second existence. That is the difference between the logical and the real world. It is these new properties that are the object of logic. You can think of indefinitely many examples of them. To be a subject, to be a predicate, to be a major term, to be a minor term, to be a middle term, to be a middle term in a syllogism of the first figure; these are so many logical properties that belong to things, not in their real but in their objective existence.

We ought then to try to rule out the confusion of beings of reason and psychological realities. I understand the lion through a disposition of my psyche (call it a concept if you please), which is something real, a psychological reality. A memory of a lion, which is simply an image by which I remember it and which can be destroyed if a hammer is suddenly applied on my skull— that is a psychological reality. I understand subject, predicate, middle term, and so on, also through psychological dispositions which are realities, just as real as anything else. The relevant point here concerns not that through which I understand but the object understood. Lion: real; deer: real; devouring: real; carnivorous: real; subject in the proposition "The lion is a carnivorous animal": that is a logical property. You see that it does not exist in the jungle. And it cannot exist anywhere else than in the mind in the capacity of object. Why? Because it is a property that things acquire as a result of the peculiarities of the second existence that they enjoy as objects of consideration, as objects of knowledge. It should be clear, then, why the possibility of making real a being of reason, a logical property, is excluded. These are properties that result from existence as objects. So, in the real world it is simply contradictory to fancy that they may exist. Those logical properties are not contradictory in themselves. There is nothing contradictory about a predicate or a subject. What would be contradictory would be the *realization* of a predicate. The day will never come when you can tell me, "I shook hands with a predicate in the street." That is impossible because it is a strict contradiction.

It is obvious that we have here a linguistic and almost a

social problem concerning the word "object." A young friend of mine who taught logic to freshmen told me that they all come to college with the interpretation of "object" as the thing that you aim at, an end, a goal, an aim. That is not astonishing at all because they are practical boys. And the object of practice and of the arts has the character of an end. So it is no wonder if object and end are lumped together in the mind of freshmen. When they are so in the mind of philosophers too it is less excusable, and it is too bad that it should happen. On the other hand, there is something much more serious, which is the identification of object with thing. Many people will tell you, "This table exists objectively," meaning thereby that if I cut my throat and go out of existence and you also and all men, the table will still exist. Now, pay attention to the role of object in all theory of knowledge, including the theory of knowledge that you are using every day, and you will see that far from meaning real, "object" means almost the opposite. For instance, there are objects in a dream, represented objects. Do they exist objectively? It is even the only way they exist. They do not exist as things, but they do exist as objects. We just have to reflect upon those things and upon our spontaneous use of words to see the difference between real existence and objective existence. I beg you to pay attention to that. Words have an awfully tyrannical power and can pervert anything.[2]

When speaking of beings of reason, the first domain to consider is obviously that of logic. Logic could be defined as the science whose object is constituted exclusively by beings of reason. Does this mean that any consideration of reality is out of place in a book of logic? That is another question. Just remember the example of the lion and the deer and it is clear enough that the logical beings of reason are grounded in reality. It is because the lion actually devours the deer that in the proposition, "The lion is a carnivorous animal," "carnivorous" is predicate. You see how the logical is grounded in the real. So far as I can see, in order to be understood, in order to be intelligible, the logician should be constantly considering the real foundation of logical properties. So, even if a book of logic is supposed to give you an understanding of logical objects, do not be surprised if it is filled with considerations relative to the real world, under either its physical, metaphysical or psychological aspects. For example, in his treatise *On Interpretation*, Aristotle considers the logical division of propositions into contingent and necessary. That involves a physics and a metaphysics of contingency and necessity. In a philosophy like that of Spinoza, if Spinoza could be absolutely consistent, *non datur contingens in natura rerum,* there is nothing contingent in reality. That is a motto of Spinoza. How consistently he lives up to that, I do not know. Suppose that a philosopher is absolutely consistent in developing a philosophy of universal and absolute necessity; for him the division of propositions into propositions whose matter is contingent and propositions whose matter is necessary would make no sense. Aristotle, however, is quite normally led apropos of this logical division to expound his philosophy of contingency, so that if you want to write a paper on contingency in Aristotle, you will have to consult not only the physical and the metaphysical writings but also his logical works. Wherever he is concerned with the division of propositions into necessary and contingent, you are likely to find some remark on necessity and contingency in the real world because that is where the logical properties of these propositions are grounded.

What happens at all times and very much in our own times, indeed, is that logic is treated the way that arithmetic was treated when I was in grade school. I must confess that when I perform a division the result may or may not be correct, but even when it is correct what I do is to recite what I was taught when I was about eight years old; a certain mental automatism was built into me then and by having it work I get a certain result which generally turns out to be, indeed, the correct one. If you ask me why I obtain the correct result by applying those rules, it would take me considerable months of work to figure it out. In fact, I do not think I ever tried to see if I understood why by applying the rules I was taught as a child I can have what is the result of the division of 233 by $27\frac{1}{2}$. On the other hand, at all times there has been a tendency to treat logic, at all levels, the way arithmetic is treated in elementary schools, that is, as a calculus which works without there being any need for understanding. In other words, we have the substitution of a possibility unintelligent calculus for an intelligent art. That certainly did not start in the nineteenth century. You notice things like that already in the thirteenth and fourteenth centuries. Take for instance the tremendously successful logic of Peter of Spain, who lived at the end of the thirteenth century; this book remained popular and widely used in schools for generations—indeed, centuries. I do not think there are any explanations in it, just rules which are perfectly sound, which work and perhaps thus provide a valid excuse for not taking the trouble of understanding. It is obvious that in our time some—I do not say all—important characteristics of that tremendous movement that is going on in logic and mathematics are traceable to the purpose of reducing those subjects to a calculus that may work without understanding. Of course, for philosophers it is heartbreaking, but you understand how those things happen. They occur under the pressure of needs which are by no means philosophic, logical or scientific, but social. In order to have our factories work, in order to have missiles detected in time, in order to stop them if they can be stopped at all, thousands of men must master forms of calculus quickly, the way we perform a multiplication or a division. If you ask them to understand, their attempts will slow down the process. Besides that, such a requirement would restrict the number of those who will be able to be accepted. Obviously, such a requirement is impossible. These historical, social necessities are perfectly respectable so long as they are known for what they are. It is not the nature of logic which demands the substitution of a calculus for the understanding of logical relations. What demands it is a certain state of society. I hope that there will always remain a few logicians dedicated not to virtuosity in calculus but to the understanding of logical properties. This is not the same as questioning the necessity of the establishment of a calculus for the operation of a very important aspect of social life.

In logic all objects are beings of reason by strict necessity. What about mathematics? Here, there are two points to be made. First of all, a mathematical entity always *may* be a being of reason in the sense that it is just as good if it is a being of reason as it is if it is not. As you know, the symbol "i" in algebra means the square root of −1. Is it a being of reason? By all means. By all definitions of number and of multiplication, the square root of a negative number, an imaginary number, is a thing not only impossible but contradictory. It is contradictory by the definition of multiplication. If you multiply 1 by +1 you get +1. And if you multiply −1 by −1, again, you get +1. So that the square root of −1 is a thing which does not and cannot exist actually, and yet it can and does exist in the capacity of object, in which capacity it plays a considerable part in mathematics. One day a colleague told me, "It is not a being of reason because it appears in the computation of my electricity bill." I think the latter is true although I do not know much about electricity and the way my bill is computed. I understand that in alternating currents the square root of a negative number plays a part. What then is it but a being of reason with a foundation in the real world? It is not an arbitrary fiction, so I have no objection to its being used in the computation of my electric bill because its relation with the real world is obvious. Now, if they brought in fairies or undines in the computation of our electric bill, we of course would refuse to pay because here there is no relation to the real world. This is the first remark, that a mathematical entity *may* be purely and simply a being of reason.

The second remark is more fundamental. If a mathematical entity is not purely and simply a being of reason, it still implies a condition of reason. What is decisive here is to understand the character of mathematical abstraction, as a result of which if and when a mathematical entity is not purely and simply a being of reason, it still implies a condition of reason. Consider a simple example. I have already mentioned it. The word "triangle" designates a gadget that you can buy in an office supply shop. It is a physical thing. If the house burns down a triangle made of plastic will burn. If it is made of celluloid it may even help the house to burn down. If it is made of metal it will melt, or at least, be distorted. However, "triangle" also designates a geometrical entity, which cannot be destroyed by fire. This entity, though relative to things capable of existence, involves a condition of reason which makes it invulnerable to fire, acid and any physical agent. So, if I think of the mathematical triangle, can I realize that thought of mine? I can take a piece of paper and cut it with a pair of scissors but the result will be a physical triangular thing made of paper, it will not be a mathematical triangle. I think we have a lot of what we need to understand the relation of mathematics to the real world if we keep in mind two considerations. When a mathematical entity is closest to the real world it still involves a condition of reason which makes it incapable of real existence; so, to realize a mathematical triangle is impossible because as soon as it exists otherwise than in the capacity of object, it loses its mathematical condition and acquires a physical one. A mathematical object, even as close to the real world as a sphere or a triangle, simply cannot be realized. Following upon this basic characteristic of mathematical abstraction, it is always possible for a mathematical entity to be purely and simply a being of reason, as in the case of the square root of a negative number, to be what is called an imaginary number.

These things have been known for quite a few centuries. They are rather clear in Plato and even clearer in Aristotle. Then they were formulated almost exactly in the terms I am using in the Middle Ages. And yet the human mind has been haunted by a vague belief that mathematics was an explanation of real quantity. That belief haunted the human mind until the non-Euclidean revolution. The great revolutions that took place in mathematics in the nineteenth century would not have shaken the human mind so badly if the implications of those extremely old propositions had been better realized. But the temptation was too great to treat mathematics as a philosophy of real quantity and you see how mechanism, especially in its Cartesian form, favored that illusion. Cartesianism is possible only if mathematics is a philosophy of real quantity. As we have seen, Descartes does not accept in physics any principles that are not accepted also in mathematics. For that proposition to hold, mathematics must be a philosophy of real quantity. So, in logic we have an object constituted exclusively by beings of reason. In mathematics we have objects necessarily modified by conditions of reason and possibly constituted by beings of reason.

Because the human mind was haunted, up until the non-Euclidean revolution, by the belief that mathematics was a theory of real quantity, it was commonly thought that the universe was Euclidean. But the real world is not Euclidean, nor is it non-Euclidean. It is a world of physical quantity, and both Euclidean and non-Euclidean geometry are dealing not with the world of real quantity but with a world of physical quantity and relations that have been already treated by a special process of abstraction. You will read in a lot of places that the non-Euclidean geometries have brought to light this sensational novelty, that the real world is not any more Euclidean than anything else. That should have been obvious long before. The day it was understood that there is such a thing as a mathematical abstraction distinct from physical abstraction, that mathematical objects as a result of the abstraction which gives birth to them are affected by a condition of reason, that an unqualified being of reason may be a very good mathematical object, that day it should have also been understood that the real world is not any more Euclidean than non-Euclidean. But of course that would have been expecting too much: freedom from imagination!

There is here a very interesting problem that arises, that of abstraction versus construction. When you hear of an activity in mathematics or elsewhere, let us say in mathematics, which is not abstraction but construction, remark that this construction takes place inside a sphere—I use an obviously metaphorical term here—into which you have entered by the basic process of mathematical abstraction. For instance, consider the properties of the square root of a negative number. Here we could speak of construction since there is nothing in the real world out of which we could at this point abstract, disengage or pull them out. I cannot take the real world and pull out of it the properties of a being of reason like the square root of a negative number. That is clear. However, notice that we enter into the sphere of the negative numbers by a process of mathematical abstraction which yields the mathematical number. It does not begin by being imaginary. It begins with what they still call real numbers. But those real numbers are already mathematically treated. They are like the triangle that I cannot make real without making it physical. Even what is called a real number in mathematics cannot be realized any more than the mathematical triangle without losing its mathematical condition. To abstract is to pull out. Sometimes when some intelligibilities have been pulled out of experience, by that very operation we have entered into a domain where many constructive operations will be possible. Of course, the combinations effected in that domain are not pulled out of the real world but it is by pulling out of the real world such mathematical concepts as number, unity, time and so on that we have entered into that domain.

I think that this question has been falsified by the empiricists of the nineteenth century. I refer especially to John Stuart Mill. For him, an abstraction was an empirical operation consisting in looking at the data of experience and shaping outlines. You look at 100 horses and then keep in mind an outline of the horse, and that is abstraction according to John Stuart Mill. I think that when mathematicians emphasize construction against abstraction, they still have in the back of their minds the notion of abstraction which was popularized by the empiricists of the eighteenth and nineteenth centuries, but especially by Mill, the most brilliant, the most learned, the most encyclopedic and the best writer of them all. It is then to be expected that you will find beings of reason in the mathematical interpretation of nature. On account of the role of beings of reason in mathematics, wherever the interpretation of nature is mathematical, you can expect a pullulation of beings of reason. In the sciences of the real, let us say in metaphysics, in the properly physical, non-mathematical interpretation of nature, in common thought, in the common interpretation of nature, do we find beings of reason? Surely. Perhaps it is not made necessary by the nature of the object, but it may be made necessary by the circumstances of thought as happens when we have to understand negations. When we treat of negations and privations and of certain relations we could not think of them without treating them after the fashion of real beings and so constructing them into beings of reason.

We have mentioned incidentally that there exist—notice that I am very conscious that I say "exist" here, and I mean it— beings of reason that have no foundation in the real world, that are constructed, indeed, out of parts taken from the real world but whose construction is arbitrary. These are not governed by any necessity inherent in the real world. Consequently, they are never scientific objects. You cannot study the physiology of chimeras. Such beings of reason may have a part to play in culture but not in scientific research. For example, a centaur is a monster which is supposed to combine the nature of the horse and that of man. Is such a combination possible? Note that if it is possible, the centaur is not a being of reason. A chimera is a monster that combines the head of a woman and a goat. Is that combination impossible? I am not so sure after all. Is it absolutely impossible to have the psychological and emotional characteristics of a woman in a physical body partly shaped like that of a goat? Try to show that it is impossible! It sounds unlikely, but that is about all we can say. A centaur and a chimera are not perfect examples of beings of reason without foundations in the real world. If by a zombi we mean a living corpse, then we are sure we have a being of reason. For then there is a contradiction embodied in that notion and that makes it a perfectly safe example. The nicest paradigm I know for a being of reason without foundation in the real world is found in the poem *Undine* by La Motte—Fouqué, a great name of the early German romanticism. It is a very lovely story based upon the myth that in the mysterious forests and swamps of romantic Germany there exists a race of beings that look and act like men and women but have no souls. That is a being of reason without foundation in the real. You see how the components of that representation are taken from the real. They are put together according to a law which is itself altogether a free construction of the mind. With such a being of reason you can write a poem as La Motte—Fouqué did, but nothing scientific can be done with it.

We may note parenthetically that it seems that sometimes the ability to manage scientific beings of reason is not unrelated to the kind of fancy which produces undines. Thus, *Alice in Wonderland* was written by a mathematician. I think that it is pretty clear that in the case of Lewis Carroll the imagination which produces undines and the imagination which produces original, creative mathematical developments are psychologically related. But this is a purely psychological side remark.

We have remarked that in mathematics we have always at least a condition of reason, and we can even have an outright being of reason as in the case of the square root of a negative number. But both in common thought and in all the sciences of the real we use not only negations and privations[3] but also relations which are of reason. For instance, if I say that a cat may look at a queen, I imply that a queen may be looked at by a cat. Now, consider the relation of the cat who is looking at the queen and the queen who is being looked at by the cat. The relation is real in the cat but it is not so in the queen. A cat who looks at a queen does not bring about a reality in the queen. The change, the new reality is entirely in the cat. That is why a cat may look at a queen. But to understand what is happening we are not afraid to say that the queen is being looked at by the cat and the relation of being looked at is a *relatio rationis,* a relation of reason. So you see that we find beings of reason everywhere, even in the most realistic interpretation of knowledge.

We have gone over a number of beings of reason and now we are back to our subject. We could not consider that subject without going over all those types of beings of reason. When we think of motion or of time now, let us leave out of the picture all of the mathematical treatment of either. Let us be concerned with physical reality. No matter how sound, no matter how fruitful and useful a mathematical treatment of it may be, it is good to know what pertains to physical knowledge, to natural knowledge, the science of nature ·qua science of nature. The problem of time, of which we shall speak a little later, is very similar to that of motion. Considering motion in as physical, as opposed to mathematical, a fashion as is possible, we have a very particular case of an intelligible system part of which is just abstracted from experience and part of which is contributed by the mind, contributed, I mean, as a nonrealizable construct, as a being of reason. We cannot understand motion, or time, without involving in the notion of either a unity which neither does nor can exist except in the mind in the capacity of object. This is extremely striking and important. This is a way to resolve many difficulties in the philosophic understanding of nature and in the understanding of the philosophies of nature.

If we are trying to understand motion, a good thing we can do is what we have done, to scrutinize Aristotle's famous definition of it. That is as good an approach as any. What we want to understand is something physical, something that belongs to nature, indeed, the center of all speculation about nature. And what do we discover? That our concept of motion and our concept of time as well, is an intelligible system in which one part is an expression of the real world and another part is purely and simply a being of reason, not of the zombi or undine type but of the type known in logic and in mathematics, a being of reason grounded in reality, one that is stabilized and subjected to laws because it is grounded in reality. If there is a component of reason in my understanding of motion and time, you can see at once that I am in permanent danger of projecting into nature something that really belongs to a construction of the intellect. Further, what is the component of reason of our concept of motion and of our concept of time? It is precisely the kind of unity that we attribute to motion and to time. So, what we are likely to project into the real world in case we are not on our guard is an unreal unity of motion and time. With it, we stop motion and we stop time; a procedure that is extremely frequent in the interpretation of nature.

With motion we have a very special case. What we have to understand is altogether real. To deny motion is to deny nature; it is to deny sense experience. And if we do that, everything is gone. We understand motion in a certain concept and as we analyze this concept we see that the fluent reality of motion is caught in a circle of unity contributed by the mind. So I have a rather exceptional case of a real thing understood in a concept whose constituents are centrally real but involve indispensably a contribution which is of reason. I do not say "a being of reason," because being designates a whole and the center and the substance of that whole is real. What is striking, what is difficult and what is supremely important is to see that the strange reality of motion or the strange reality of time can be understood only in an expression part of which is of reason. What is contributed by reason is the togetherness of the before and the after. The before and the after exist unqualifiedly. If you are a geologist, for instance, and study strata, you can determine with a high degree of assurance that a certain geological stratum was much earlier than another. If there had never been any geologists there would have been a before and an after all the same. The point is that the unity of the before and the after is what is contributed by the reason. The arrow that flies from here to there is flying through here, indeed, before it flies to there, whether there is a philosopher to watch it or not. What calls for a contribution of the reason is the holding together of the before, which no longer is, and the after, which is not yet.

Several decades ago an eccentric philosopher, whose name I will not mention, wrote an impressive note on the nature of motion centered on this remark that the notion of motion is not entirely contributed by reality but involves a contribution of the reason. In the last sentence of his paper he explained that what he had to say could not be accepted by his contemporaries, so that he was really writing for a future generation. What he claimed is that Aristotle's theory of motion purely and simply renders nature to a large extent subjective. It would really be queer if Aristotle of all people had initiated a theory which renders nature to a large extent subjective. That would be a really new version of Aristotle. He has already been interpreted in a thousand ways but that would be indeed the most sensational interpretation of all. But, that is what our man wanted to say because obviously he had a subjectivistic philosophy to sell. I do not mean that it was dishonest, of course. People do those things honestly. But there is absolutely no subjectivisation of nature in Aristotle's theory concerning the concept of motion or the concept of time or the concept of counting angels or judgments.

If I say that when I count judgments the concept of a number of judgments is not expressive of reality without also being expressive of a contribution of the mind, it does not mean that Peter, Paul, and Bartholomew do not really have their own judgments. This does not do anything toward the subjectivisation of nature. That would be a complete misunderstanding. The being of reason part of the concept is not here to make nature subjective. On the contrary, its function is to render nature, such as it is, intelligible, to make it possible for nature to appear as an intelligible object. Of course, in order for the reality of nature to be understood with such a concept we have to distinguish sharply what comes from nature and what from the mind. If we mix up the two, we are wrong, though it is a frequent accident, in particular in the interpretation of time.

It is an extremely common accident to substitute for time something that has a unity incompatible with the nature of time. In other words, we take our concept of time with the unity contributed by the reason, we drop the physical content and then we have something which is purely and simply a being of reason but which we still call time. That happens often indeed, and I think it happens normally in mathematical physics. This procedure becomes vicious when this being of reason of mathematical physics is held to be purely and simply expressive of reality. As a result, you have philosophers of time and becoming whose work is a protest against such a reduction of time to what it is not, a reduction which takes place along the line of a unity contributed by the mind. The philosopher I have in mind as most typical of this is Bergson. That is what he wrote about primarily. He began his career as a mechanist, then one day understood with all the forcefulness of his philosophical genius that the time of physics was not real time. He did not say that physics was bad. It may be entirely normal and sound in a certain discipline to work on a being of reason related to real time but impossible to identify with real time. He did not say that physics was bad, but asked all his life what physics was doing and what should be done in order to recover the reality of time and with the reality of time the distinction of things psychological and things physical. That is the framework of the work of Henri Bergson. That is also the reason he is important in the history of ideas.

NOTES TO CHAPTER FIVE

1. *The Material Logic of John of St. Thomas,* trans. by Yves R. Simon, John J. Glanville and G. Donald Hollenhorst (Chicago: University of Chicago Press, 1955).

2. In everyday English "thing" and "object" are synonymous. In epistemology "object" is opposed to "subject" and a thing need not be an object, nor an object a thing. A thing becomes an object as it becomes known or knowable. Thus, to be an object (objectivity) always involves a relation to a subject. So, what I dream about is an object and exists objectively, that is, as something being known by me in my dream, but it need not be a thing that exists outside of my mind. To put it in another way, an object is that with which an operation, be it cognitive or appetitive, is concerned. For a full discussion of the meaning of object and subject, see L.M. Regis, *Epistemology* (New York: Macmillan, 1959), especially pp. 175–252.

3. A negation is a lack of being which we think of as though it were something, e.g., a whole. A privation is a negation of that which a subject ought to have, e.g., blindness, for a normal man has sight.

Chapter VI

PLACE AND SPACE

We have seen that for centuries there has been taking place a dialogue between those who interpret the world in terms of natures and those who interpret it in terms of space. Without doubt, the main characters of the dialogue are Aristotle and Descartes. We saw that very well when we studied their views on motion. Now concerning the subject of place, we find that the discussion of it is simply an extension of the Aristotelico-Cartesian debate. For the time being, however, let us leave aside the question of the nature of the relation between place and space. One thing I can say right away is that the character who represents nature will speak of place first. And if he does speak of space, he will speak of it only later. But the mechanist, the character who impersonates space, will speak of space first and if he speaks of place at all, it will be later and in reference to a pre-established concept of space.

We must from the beginning be on our guard against taking the concept of space for granted as a primitive one. We realize that this is a very common practice in modern philosophy. How could it be differently, modern philosophy having been founded in some parts of the world by Descartes and in other parts of the world by Hobbes? Wherever in particular the Cartesian influence is decisive, space will have a character of radical primitiveness. Thus, the substance of Spinoza unfolds itself according to two modes which are thought and extension. On the other hand in Leibniz we have a problem. I would not be too categorical concerning the primitiveness of space in Leibniz. I

113

would say nothing one way or the other except that one who would grasp the real status of space in Leibniz would perhaps have understood Leibniz. I think that nobody ever has so far. When you come to the great idealists like Berkeley for whom material things do not exist except as objects of thought, what about the status of space in that world of things whose to be is to be perceived? I think that there also it is primitive. But then we have Kant. The concept of space has a very strongly marked character of primitiveness in the philosophy of Kant. And then when you take, in the later nineteenth century, those combinations of Cartesian rationalism and Kantian formalism which have one taste in Germany and another taste in England and a third taste in France and in the countries of French influence, what matters is the dual primitiveness of thought and space.

Space is not primitive in Aristotle. Whether it can be primitive in a philosophy which interprets the world in terms of natures remains to be seen.

Let us begin by first going to the fourth book of Aristotle's *Physics*. What we call place, he calls *topos*, the word from which we derive topology and a few other more or less scientific terms. I beg you by the way not to identify, as a few great thinkers did, without any further ado what is called *topos* in the *Physics* and what is called *pou* in the *Categories*. *Topos* is place. In the *Categories* what is called *pou* means properly "where." The categories of Aristotle are substance, quantity, quality, relation, action and passion, when, where, posture and situs. The category *where* is not place; it is the accident which results in a substance from its being in a place. Take the predicates of the propositions: "He is in his bedroom," "He is in the market place." Those predicates belong to the category of *where*. They do not designate a place; they designate that which results in a subject, for instance, a man, from the fact that he is in such a place or in a different place. You will find histories of philosophy (sometimes done by competent people, which should lead you to take competence with a grain of salt) in which these two categories *where* and *when* are cold-bloodedly translated as place and time. That is definitely wrong.

Concerning "place," we may say the great theory is that of Aristotle for the other theories are really concerned with something else. He is *the* philosopher who really attempted to answer the question: "What is place?" Considering that all things have their own place, you may raise the question: "Where are they?" and define criteria for answering it. But you may also ask, since we say that they are somewhere and that they are in their places: "What is place?" And that is what Aristotle tried to answer.

I call your attention to the character of the fact which is decisive and at all times guiding Aristotle's inquiry into the nature of place: it is the fact of *antimetastasis,* that is, the fact that one and the same place can be occupied *now* by one body and later by another. This is a very striking example of a philosophical fact—of the kind of fact that the philosopher uses. Much can be derived from the consideration that when a bottle of wine is emptied it is simultaneously filled with air: an instant before a liquid, and now a gas; and again you can pour water or wine into it, air will go out and be replaced by the water or wine. This is the fact which for Aristotle establishes the reality of place and its intrinsic character.

Think of an example such as this when you reflect on the great subject of the relation of philosophy to experience in order to determine the nature of the experience which is decisive in philosophy. At the turn of the century there was a widely accepted fashion of founding philosophy on scientific facts. I have known that fad. The facts had to be established by the sciences. You took the facts as established by the sciences and then you began to philosophize. I was fed that epistemology as a young man. It upset me deeply—which was not a bad thing of course. It moved me to acquire just about as much scientific instruction as I could. It was not much but it was better than nothing. For a while, at least, my great motivation was the fear of lacking an experiential basis if I did not have plenty of scientifically established facts. In the work of Bergson there are a couple of passages, really beautiful ones, showing that if you take scientific knowledge you also take the forms of scientific knowledge and you will never reach philosophy. Or, more exactly, you will substitute for philosophy a pseudo-philosophy embodied in habits of the scientific mind. It is not necessary to follow the philosophy of Bergson in its peculiarities to see how exactly he put his finger on this point.[1] If, for instance, you take facts concerning electrons in their scientific formulation, no matter how good and well-established they are, you remain in physics. And for such facts to be of philosophic use a philosophic reinterpretation has to take place and it is not always possible. In other words, no matter how empirical our philosophy it cannot be constructed on facts that are anything else than philosophic facts. That is what is important, the notion of philosophic fact. There are plenty of facts that have a philosophic character. For instance, being exists; plurality exists; things are not immobile but some of them at least are in motion. These are philosophic facts. And perhaps if our philosophy were really well organized we would be able to point to the philosophic fact which is the experiential foundation for every particular chapter or development.

Aristotle considers and defines place in the fourth book of the *Physics*. Roughly, this is how he proceeds. Consider a certain subject, it might be a stone, but let us make it a man for ease of expression. When we say that a man is, for instance, in the United States or in the state of Illinois or in the city of Chicago or on the campus of the University of Chicago or in the Social Science hall, we are answering the question of where he is, but we are also expressing a place with an increasing precision, which is still far from being absolute. Here is what I think is decisive. Aristotle from the beginning envisages place as a container. Where is that other man? He is contained in . . . some place called Mexico. That is vague, certainly. Is it a tridimensional container? Well, no doubt. Mexico contains not only plains and mountains but also at least a part of the air on top of the plain and the mountains. Granted that, is he in Veracruz or Pueblo, or Mexico City? This we could leave indeterminate, but let us say he is in Mexico City, in the Plaza de la Constitución, in a hotel, in the dining room, in the northeast corner of the dining room—what are we doing? We are bringing out determination of the container closer to the contained. Every time the determination of the container is brought closer to the contained, the question "In what place is it?" is better answered. "Better" here means, as it normally does in the human knowledge which is progressive, less vague, more precise and more determinate. Now, you cannot squeeze the container indefinitely, otherwise, the thing contained within it would be crushed. So, there is a final step that is perhaps extremely difficult to effect, but you have to consider it. It is the final step in this narrowing down of the container. Where is the good wine of which you spoke to me? It is in our country house. It is in the basement of the country house. It is in the northeast corner of the basement. It is in a locker of which I have the key. It is in the upper part of the locker. It is in its bottle. Now, the bottle is about ¼ of an inch thick—that is not too thick for such good wine. So, in order to finish the determination of the container, you still have to do what Aristotle does, to consider the ultimate container, which is the inner surface of the bottle. That is where the wine is. The example of wine and bottle is in Aristotle. That is very simple and very plausible, is it not? Indeed, that is the way we always speak, is it not? To tell where someone is, we do not need to come up to the absolutely ultimate container but we certainly have to try to narrow the container down somewhat. For practical purposes it is not generally necessary to reach the last limit, but we understand that the notion of container and the narrowing down of the container would make no sense if we did not bear in mind this ultimate limit. I think this very important. Unconsciously—those things do not need to be conscious—as we are expressing the vaguest notion of a container or, more definitely, as we are gradually determining and narrowing down the container, we are guided by this absolute, the absolute container which is the inner limit of the last container, defined as last by the contiguity of the thing placed and this very last limit.

All that is in Aristotle, but he does more than that; he includes something else in his definition; the motionlessness or immobility of this last limit. The thing placed here is placed in a container that Aristotle understands as motionless. What he means is very simple. Take the bottle of wine. We like nothing better than to empty it into our glasses. But then what happens? No more wine in the bottle—just air. Thus, the inner limit of the container is motionless. Why? Precisely because of the fact of mutual replacement; the liquid gets out of the bottle and air gets into it. We say that it is the same place which is emptied of liquid and filled with air, but we could not say that it is the same place if that place were not, as place, considered free from motion.

The theory of place in Aristotle then is absolutely clear. The philosophic fact which is the empirical foundation of the theory is the experience of a change of substance within one and the same place. It takes no more than this: pour wine into a glass, but simultaneously notice that the inner limit of the container which was the place of wine is now the place of air. The wine has changed; the place remained itself, unaffected by the change. So, it all ends in this classic, well-known definition which you find in the *Physics:* place is the motionless, immediate limit of the containing body.[2] And this is obviously a result of Aristotle's concern to interpret the world in terms of nature.

This motionlessness of place, the including of immobility in the definition of place, causes obvious trouble; we only have to think for instance of an object which is contained in a container that is itself in motion. Think of a boat anchored to a tree in a river. If you ask where the boat is, I can answer with great precision by defining relations to a given tree, to the top of the hill, to the city-hall, to several poles on the banks of the river. But if you ask the question: "What is *the place* in which the boat is?"—and that is precisely the question that Aristotle would like to answer—you are implying that the boat is in a place; and what Aristotle struggles to do is precisely to define the nature of *the thing* that we assume when we speak of place. If we ask *"where* is that thing?" or "Where is the motionless container in the case of a boat anchored in a river?" we may answer that it is the river itself. But then we realize that our answer is vague; it is as if we said that the good wine is somewhere in the neighborhood. Of course the neighborhood and the river are not that ultimate internal limit of the container, which was given as the definition of place.

This is a difficulty that does not seem to be answerable from the point of view of Aristotle and was big with consequences in the evolution of physical ideas. We notice very early in this development of scientific ideas how there is substituted, for the consideration of place, a new sort of response to the question of where a body is: the answer is given in terms of relations. Of place, there is no more question, or, if the word is still employed, its meaning changes; it becomes what we correctly call space.

Let us then consider briefly a problem which must certainly be bothering you, the relation between place and space. Of course, the word space has been and is often used very loosely, and to try to formulate even a nominal definition might perhaps seem rash. However, it seems to be pretty safe to declare that space designates a distance. Somewhat paradoxically, the way to speak of a space of time shows this. What do we mean when we speak of a space of time? We imagine two points in the running of time and we consider a distance between these two points. The idea of a distance is common to the most different interpretations of space. So you see that we are not dealing with a primary notion. Before we work out a notion of space we need to have bodies; these bodies have to be external to each other and placed or located apart from each other. We may then consider a point there and a point there and work out the notion of a distance beween these two points.

The thing of great importance for the whole history of the question is to realize that the system of distances—generally tri-dimensional—which is called space may be real, but it may be imaginary just as well. That is, what we mean by space may be something real, but it may just as well be a being of reason. Keep that in mind when you read about the evolution of physical ideas, especially concerning this particular subject. There is nothing to prevent us from constructing an imaginary space, placing things in that system of distances and, of course, it will be exceedingly easy to treat this imaginary space, this imaginary system of distances, as if it were something real. When we surveyed the theory of the beings of reason I mentioned that in mathematics, and by virtue of the very nature of mathematical abstraction, a being of reason may always be just as good as a real being. There are theories in which a real number is considered just a particular case of a general notion of number which may be indifferently real or imaginary. What holds in mathematics holds also for the mathematical interpretation of nature which was not invented in the seventeenth century but is very old, having received considerable development, so far as astronomy is concerned, at the dawn of Greek civilization. Now, suppose that we are astronomers concerned with a mathematical reading of sensible appearances: using the notion of an all-containing space, of a body which may be indifferently full or empty and in which everything takes place, will be most natural from our standpoint since we shall be trying to interpret the physical world with mathematical instruments. This is what happened early in the evolution of ideas. See Proclus [3] for instance. In the astronomical, mathematical and philosophical writings of Proclus the word *topos* has come to designate an indifferent body which contains all things. Does he hold that this body is real? I doubt that the question is decided. What is important is that, whether there are bodies in it or not, it is there ready to contain whatever is brought into existence. It is space in the modern sense. The big issue: is it considered as something real or is it an imaginary system of distances where everything can take place? From an Aristotelian standpoint real space is nothing else than a system of distances between definite points in existent bodies—anything else is a being of reason, grounded indeed in the real world, admitting of scientific treatment and capable of doing a lot of good work in the mathematical interpretation of nature. Keep that in mind when you study the notion of space in Leibnitz for instance, or in Kant.

When I consider the subject of space in modern philosophy I think first of Leibnitz and Kant, and of all the idealistic trends that derive from them; I think particularly of those many trends in the nineteenth century which took completely for granted the idea of forms of space and time as defined by Kant. In reading his "Transcendental Esthetic" never forget this problem of what we call space, this indefinite container in which our imagination places things so handily. For although there is no reason why it should not be a being of reason, there is such a thing as a real space. And we do not need to trace it, in Kantian fashion, to the nature of our mind and to *a priori* conditions of sense perception. On the other hand, if we have clear notions about the origin and the operations of beings of reason we may get quite new and fruitful insights into the work of Kant and generally into the system of problems which has been worked out as a result of his ideas.

One very serious issue in the system of Aristotle is the notion of natural place. I have been alluding to it a number of times. I was bearing it in mind in particular when, at the beginning, I drew a comparison between motion as understood in a philosophy of nature and motion as understood in mechanism: the motion of Aristotle and the motion of Descartes. Permit me to repeat once more something very important: the motion of Descartes has been already treated mathematically. How can a motion which has been already treated mathematically coincide with physical reality? That is one of the secrets of the philosophy of Descartes. If we want to get to the core of his philosophy, that is the kind of questions we should ask. Anyway, we recalled that the first thing which happens when a subject is treated mathematically is the disappearance of final causes. There is no final causality, no goodness, no desirability in mathematical entities. Thus, for Descartes no object has a natural place.

In Aristotle, on the contrary, motion is something physical and in no way reducible to the motion of the mathematicians, that very primary mathematical notion which is antecedent to the notion of surface and to that of line. Rather, motion is an act, a certain kind of act, the act of a subject in potency considered precisely as such. Wherever there is act in Aristotle, it involves finality. For, act is the end of potency; the imperfect actuality of motion is relative to the perfect or complete actuality of the state which terminates motion. We can think of examples in which it is perfectly clear: consider the growth of a living organism for instance. This is an instance of the motion called increase. There is a certain size which is normal for an adult organism; watch the young animal or the young plant developing towards that normal size and the teleological character of the process is quite clear. If that plant or animal or especially a member of the human species is stopped in that development, we consider that it is too bad: a certain state having the character of a good has not been reached. This teleological aspect is rather clear in some cases of increase or decrease and of qualitative change. In cases of local motion it is much less clear, especially after our minds have been conditioned as they are by the mechanism of diverse modern philosophies and by the mechanism which is the method of modern science.

What does Aristotle do about teleology in the case of local motion? Roughly, here are his ideas. The world is made of a very few elements. The tradition of the four elements and of the elementary qualities is, I think, as old as mankind and as general as mankind itself—you find it in Greece, in India, more or less everywhere. You can see why the notion of elements is so interesting and so universally appealing to the human mind; it ties in so well with the universal endeavor to cut the reality of change to a minimum, to find something permanent under changing appearances. One of the things we do in order to overcome the bewilderment of change is to hold that diverse compounds are made of the same elementary bodies, the fewer the better. Until recently, they were really few: they were four. Those four had elementary qualities attached to them: earth being cold and dry, fire being hot and dry, water humid and cold, air light and dry. Anyway, what matters for our subject is that, in Aristotle, everything, by reason of its elementary composition either goes down or up: it is normal, it is good that it should do so! There are things that are heavy by reason of their elementary composition and their natural place is down. True, you can throw a stone up: this is *violent* motion, an energy from outside is impressed upon the stone and so the stone goes up, but remove any obstacle and the stone will go down. You can also fill a bag with air and hold it at the bottom of the lake with a stone, the "light" will be kept down: remove the stone and the "light" will go up. I do not need to say it: this is one of the points of the physics of Aristotle that has been completely wiped away by modern physics. Yet, is there any acceptable meaning to the notion of natural place? First, what does it actually mean in Aristotle? Does Aristotle mean that it is good for the stone to be down, for air to be up? I do not think so! Rather, his teleology of places is relative not to things but to the universe, to the world considered as a work of art.

Here we are going into fields that are, even historically, very obscure. As I mentioned earlier, there is now a very strong suspicion that, at a certain time of his development, Aristotle held a vision of the world which had been picked up by the Stoics, and in that vision of the world the universe has the unity of an individual. Let us consider briefly this historical problem. Old Stoicism (c. 304–205 B.C.) developed very soon after the death of Aristotle (322 B.C.). It was shaped within a generation after the death of Aristotle. By Old Stoicism I mean the philosophy of principally three men: Zeno, Cleanthes, Chrysippus. Chrysippus, the most productive of them, is credited with having written some seven hundred books, but most unfortunately only a few fragments were left, though enough to give a good idea of what these men were thinking. They were very great geniuses; we cannot overdo their importance in the history of ideas. (Many things done in logic today originate in Old Stoicism and in particular in the work of Chrysippus. If we understand well the meaning of logic in Old Stoicism perhaps we will understand what contemporary logic is about.) At the time of the Old Stoics the works of Aristotle were not available. You know of the report that the manuscripts of

Aristotle, in as much as their entirety is envisioned, were put in a basement where they stayed for two centuries, until the compilation which we know was made. During these two centuries Old Stoicism had been founded and had produced its work. Thus, the Old Stoics did not know the Aristotle we know and apparently they knew an Aristotle that is lost to us.

According to erudite research of the last generation, the Stoics found in that lost Aristotle a cosmology which they made their own. That is very important. Assuming that Aristotle, between middle age and full maturity, changed his physics, it is more than likely that in his treatises which are known to us and are for the most part the work of his maturity, there are remnants, survivals of that preceding period. So, by considering the Stoic universe, we may make reasonable guesses concerning this early Aristotelian philosophy of nature, and if these guesses are reasonable they may be helpful in understanding some features that are held over in the period of his full maturity. What I have in mind is this: for the Stoics the universe has the unity of an individual; it is like a big organism, whose soul is God. There is an intelligent divine power immanent in this one organism and so chance is completely excluded from the Stoic universe. How could there be chance in it since the parts of the universe are unified like the organs of a perfect organism? I say a perfect organism because we have come to understand that organisms have not only a nature but also a history, and if we study the parts of an animal, like man, for instance, we find a few things which seem to have no function and whose presence is accountable for by the history of this organism: in all higher organisms there seems to be leftovers of anterior organic forms; that is even one of the most striking arguments in favor of evolution. The Stoic universe is more perfectly unified than is any organism according to today's understanding of organisms, and the behavior of things becomes the behaviour of the parts of a completely unified whole, a whole which is wisely unified by an immanent intellect constantly described as providence. The word *pronoia* (Greek for providence) appears in Plato, but the notion of providence was popularized by the Stoics. I think that Christian writers, in many cases, took over the notion of providence which had been worked out by the Stoics. Of course they corrected it because, in Stoicism, this providence is pantheistic: it is immanent in the world.

If you go along with this supposition that the Stoics got their cosmology from an early version of Aristotelianism, then the notion of natural motion makes a lot of sense. Suppose that the universe of Aristotle is the Stoic universe; if this is true, then the teleology of local motion becomes inescapable. Asking about the finality of local motion in such a universe is like asking whether it is good for an organism that food should go down the esophagus to the stomach rather than up. The answer is rather obvious in the case of an organism, and if the world is attributed the unity of an organism then the notion of natural motion and the teleological character of motion becomes perfectly intelligible. This therefore is my suggestion: When Aristotle speaks of natural motion, the teleology involved does not concern merely the thing in motion but the whole of which it is a part. For the universe, it is good that the stone should fall down and that fire should go up. Suppose that the universe has the unity of a perfect organism, then there is no chance: this is not Aristotelian—it is Stoic but it may have been Aristotelian in the early stages of Aristotle's development. Suppose a universe that has the unity of a perfect organism, then every motion is good in some way, good for this whole of which the moving thing is a part. But how general is this teleology of motion? If the universe were a Stoic one, it would be all-embracing; every motion would have a character of goodness for the whole of which the moving thing is part. But, in the mature phase of Aristotle, the universe is not so completely unified: there are real irreducible pluralities and chance is real.[4] If chance is real and if the universe does not have the unity of a perfect organism, can it be said that every motion is either natural or violent, i.e., impressed in spite of nature? That is very uncertain. But it is a good thing that recent research on the remnants of an antecedent period of Aristotle has introduced greater sublety in our examination of the theory of natural places.

To conclude, I have tried to show how Aristotle resolved the problem of the nature of place. He answered this question in terms of containers that are closer and closer to the thing contained. Place, absolutely understood, is the motionless inner limit of the ultimate container. The history of physics shows us how this definition of place gave much trouble to astronomers —astronomers being the representatives down to the seventeenth century of what we now call physics. According to Aristotle, place is a motionless limit; what do we do in the case when a body is contained in an environment itself in motion? The classical example here is a boat anchored to a tree in a river. The immediate container of the boat is a fluid, a thing in motion. We can specify the location of the boat very clearly through a system of relations; by relating the boat to the tree or to the top of a mountain or to the tops of two or three mountains we can very precisely pinpoint where the boat is. This procedure however, provides, in terms of Aristotle's philosophy of nature, no clear answer to the question of what kind of thing is the place of the boat. Rather, this question is answered in terms of spatial relations. Of place, there is no more question, or, if the word is still used, its meaning changes: it becomes what we correctly call space, a system of distances which is grounded in reality indeed, but is extended in all directions by our imagination and within which it is always possible to locate a body by defining relations to fixed points or to points that are assumed to be fixed.

With this conception of space and this relational interpretation of the question of where an object is, the philosophical problem of the nature of place is left out—the problem precisely that Aristotle had wanted to treat. Thus, we can say his definition is successful when the container is motionless, but uncertain when the container is in motion.

Is it then possible to define place philosophically in all cases? Is it possible to give a definition of place when the container is itself in motion? That is the question I would raise but to which I do not wish to try to give an answer. So far as I am concerned, it is enough that we should have the difficulty in mind: I have no means to solve it.

NOTES TO CHAPTER SIX

1. For a more detailed elaboration of this point, see Chapter Eight, Section Three.

2. *Physics,* 212a 20.

3. Proclus: 410–485, one of the last of the Neoplatonists and the immediate predecessor to Simplicius whose commentary on Aristotle's *On the Heavens* is important because by comparing the text and the commentary of one of the last Neoplationists one can see the difference between an Aristotelian astronomy and what we could probably call with propriety a Platonic astronomy.

4. See Book Three of the *Physics* of Aristotle.

Readings

M. Jammer, *Concepts of Space* (Cambridge, Mass.: Harvard University Press, 1954).

D. Nys, *Cosmology* (Milwaukee: Bruce, 1942), Vol. 2, Part III.

V.E. Smith, *Philosophical Physics* (New York: Harper, 1950).

Chapter VII

TIME

St. Augustine is famous for having said that till somebody asks him what time is he knows very well what it is, but he ceases to know as soon as anybody raises the question. How to organize an exposition on this tremendous question is, by itself, an extreme difficulty. I suppose that the best is to proceed historically and, since the first organized treatment of time is found in Aristotle, I wish to take a glance with you at those basic texts of Aristotle on the subject. There is much reflection on time in the work of Plato—in the *Timaeus* in particular—but an organized and systematic treatment of the subject of time appears first in Book IV of Aristotle's *Physics* and these chapters will remain the basic text for philosophers of all times. Let us consider the successive steps he takes in this exposition.

The first thing he does is to express the connection of time and motion. Here the word he generally uses for "change" is not *kinēsis*, motion, but *metabolē* which means mutation but also means change in the broadest sense, with always an eye on the kind of change which enjoys primacy. Local motion enjoys primacy because of its continuity and, when we come to the subject of time, continuity becomes particularly significant. So more than ever, even though he is using a very broad term which can designate any kind of motion, it is the distinguished case of local motion that Aristotle has particularly in sight. There is no time without motion. How do we know that? The connection of time with motion is established for him by a psychological experience. Where we have no experience of motion, neither do we have an experience of time. Imagine circumstances of darkness, of silence and of rest that are such that, without being completely asleep, we have no experience of motion, we do not then have any experience of time either. It is this connection of the psychological experience of motion and the psychological experience of time which, for him, establishes the connection of time and motion.

Having established that there is a connection between them and considering on the other hand that the two are not identical, since it cannot be said purely and simply that time is the same as motion, Aristotle raises a very precise question. Since time is connected with motion but is not motion, it is something pertaining to motion. What then is that something of motion that we call time?

His third step consists in recalling that motion is connected with magnitude. This is because, as we saw, here more than ever by motion or change he means primarily motion in place, local motion, which involves magnitude, so that time also is connected with magnitude.

Fourth step: in magnitude, more precisely in place, there is before and after. You probably know already the definition of time that Aristotle is going to reach. In that definition the words before and after are included. It has often been asked whether we have a circular definition, before and after seeming to imply an already established notion of time. This is excluded by the order of the steps which make up Aristotle's inquiry. The before and the after which are mentioned in the definition of time concern magnitude and more precisely place: in place, it is always possible to determine an origin and an end. Considering a body in motion it is always possible to define a point as being before an other, before, not in time since we have not yet the notion of time, but in place, in magnitude.

Then we may consider this a fifth step in Aristotle's research: the before and the after of which you speak when you speak in terms of time are really identical with motion, however, they constitute a distinct object of understanding. What Aristotle means is that motion *qua* motion, motion considered precisely as motion, is the act of a subject in potency. In this act of a subject in potency you distinguish a before and an after. In reality, in the thing itself this before and this after are identical with motion, yet they are understood differently, and so constitute a distinct object of thought.

In the sixth step of his inquiry, Aristotle comes back to the fundamental experience of the connection of motion and time. When we perceive no motion, we perceive no time; when we perceive motion, we perceive time. *How* do we perceive time when we perceive motion? We perceive time by distinguishing in motion the before and the after. This perception of time takes place when we perceive two terms that are distinct and that comprise an intermediary. And this is how we reach the famous definition of time as the number of motion in respect to the before and the after.

In the following chapters Aristotle raises the question of the reality of time—a question that we are pretty well equipped to consider after all we have seen about those realities which are not understood and are not intelligible without a contribution of the intellect. Motion and time are the best known of them. In fact, time evidences more clearly than motion the contribution of the knower. But the problem of the reality of time is obvious. We divide time into past, present and future. Of the past we say that it no longer is, of the future that it does not yet exist, of the present that it is an instant. What kind of reality can we attribute to a thing made of what does not yet exist, of what no longer exists and of what has a character of instantaneity, a character opposed to that before and that after, which is in the definition of time?

No doubt, more clearly here than in the case of motion, we have to deal with a being of reason interpretative of something real. There is something real—whose nature we shall try to specify shortly—which is treated by the reason according to a system of unity that is not in the real. In the real, what does exist is motion and its phases. There is really a before and an after, but what does not exist in the real is the unity of the before and the after. The numeration or measure is effected by the mind.

We have remarked that both in motion and in time there is a unity contributed by the mind. Without that unity we could not grasp what is real in motion and time. Could we say the unity contributed in either case is the same so that the concept of time and motion would become identical? In terms of Aristotle's analyses I would say the difference is that in the case of time the unity is directly and properly relative to the before and the after, whereas in the case of motion the unity contributed by the mind is relative to the state, if I dare say, of a being that is actualized. Let me explain myself better. What the concept of motion expresses is the fluent actuality of a thing in potency—not just the actuality of an imperfect subject that remains in a state of imperfection but the actuality of an imperfect subject which, as it enjoys this act, tends toward further actuation. It is that succession, that continuous succession of actualizations which is unified by the mind in the concept of motion. On the other hand, in the concept of time the relation is not to the succession of actualizations but to the before and after in place. If we take time as a measure, I think it can be said purely and simply to be a being of reason, with a foundation in the real of course. So when Aristotle defines time as the measure of movement according to the before and the after he is directly defining time as a construct, as a work of the reason, interpretative of something real. Now, if we consider the something real which is interpreted by the measure called time, then we come to consider time not as measure but as duration. But what do we mean by duration? Is duration exactly the same as existence? Obviously not. The difference between to exist and to endure is slight, but it is decisive. When we think of enduring we include in our consideration a reference to a possibility of existence being followed by non-existence. That is how I would relate the concept of existence and the concept of duration: in the concept of duration there is a thing which is not in the concept of existence, namely, a reference to the possibility of existence being followed by non-existence. Duration is continued existence; duration is existence conceived as resisting (I speak metaphorically—but what else can I do?), as resisting the conceived possibility of non-existence.

This becomes a little clearer if we consider our experience of the two basic ways of duration. They are both given in our experience. We experience such a thing as duration by way of motion, and we also find in our experience, much less conspicuously but no less importantly, samples of duration *not* by way of motion. We have already discussed this subject when we were treating of motion. I called your attention to the insistence with which Aristotle distinguishes acts that are motion and acts that are pure activities. Of the former we have a number of examples: to hammer, to chisel, to shape a piece of clay into a statue. Any of those actions are of such nature that they are purely and simply identical with the becoming, the motion of a subject and, when this becoming, this motion of a subject comes to an end, activity also comes to an end. In opposition to this we realize that an activity such as contemplation is capable of enduring without any change having to take place by intrinsic necessity.

It goes without saying that any activity which endures *not* by way of motion is narrowly circumscribed in human existence. According to Aristotle, the good life, the best life is made of contemplation. How does he conceive of the really happy man, the man who has a really happy life? He imagines him as a person who is lucky enough to have good health, who has enough means to enjoy economic independence, who has achieved control of his passions, who is at peace also with society (he must be at least forty too, for all that does not happen with a young fellow) and who, having satisfied all these conditions, can dedicate himself to the life of study. Now, in the life of study there is a great deal of research, and so long as we are researching the best life is not yet attained. The best life is attained when research has gotten somewhere; it certainly need not have exhausted any subject, but concerning a question it must have obtained a well-established answer and then, no matter how inexhaustively we possess truth, looking at it is the cause of the loftiest joy and it is the core of the really good life in the eudaemonism of Aristotle.

It also goes without saying that such an achievement does not happen often. Consider this poor fellow who has taken the trouble of making peace with himself and with society while accumulating enough wealth in order to be economically independent, and who has been searching so thoroughly and so successfully to obtain truth in a few subjects. It is clear that when he begins to contemplate the truth that he has obtained at long last, he is performing an activity which endures not by way of motion. But within five minutes he will have to answer a call, he will need a drink, he will find it necessary to scratch because although we assume he is in good health his health is not so perfect as to be immune, for instance, to the itching of a mosquito bite or any thing of the kind. In terms of volume, things that endure without motion are extremely little in our life. Many thinkers indeed—not exclusively but perhaps especially among modern utilitarians and pragmatists—are simply unaware of the reality and of the meaning of those activities which endure *not* by way of motion. Those activities are real and they are qualitatively supreme, but we realize that, no matter how qualitatively significant and real in our life, they are just short-lived participations in eternal life, surrounded and restricted on all sides by realities that endure by way of motion. Thus, it is by comparing duration *not* by way of motion and duration by way of motion that we can get our best understanding of the reality of time.

If you consider time as measure, then, no doubt, as Aristotle puts it, time would not exist if the soul did not exist; it is a being of reason interpretative of something real. It takes a mind to effect the measure or numeration according to the before and the after. But what is the reality of time that we interpret in those terms of measure? It is the reality of an impermanent existence, of an existence by way of motion. But such a notion becomes clear only when it is related to some understanding of duration existing not by way of motion. Thus, we have here to deal once more with the problem of analogical thinking. The concept of duration is indeed an analogical one. It is not possible to abstract from time and eternity features that would be simply and purely common. Just as it is impossible to abstract a concept of being from the diverse sorts of being, so it is impossible to abstract a concept of duration from the diverse ways of enduring. Consider duration as divisible into time and eternity. I insist that this division is possible at the level of our human experience, without having gone beyond the experiences of this world and without having inquired into the existence of eternal beings. Indeed, we find in our experience such things as contemplation, joy and some forms of love which, though surrounded and restricted by motion and time, do not imply motion and time by intrinsic necessity. If you consider what a carpenter does, it implies motion and time by intrinsic necessity; if you consider what the contemplative mind is doing, it simply does not imply motion and time by intrinsic necessity; on the contrary, it rather excludes motion and time—here, motion and time are implied not by the nature of contemplation but by the circumstances. If we consider time and eternity as modes of duration we have here a splendid example of an analogical set:

E—Here you have an unqualified assertion of duration

T—Here you have an assertion of it which is coupled with a negation.

To the best of my knowledge this is the most illuminating possible insight into the nature of analogical unity. When we have to deal with a univocal concept, there is a common ground which is asserted in all cases; for instance, if I say that a horse is an animal and that a camel is an animal the difference between a horse and a camel may be considerable, but there is something common which is expressed by the word "animal." That something common is asserted in the case of the horse and in the case of the camel; it is not affected by any particular negation in either case while, when you consider the analogical concept of duration as divided into time and eternity, the common ground expressed by the word "duration" is asserted purely and simply in the case of eternity and it is both asserted and denied in the case of time. For, to exist in time is to exist by way of motion, it is to retain existence at the intrinsic and necessary cost of a loss.

I do not know if this is quite clear. Sometimes it happens that it is in art and in poetry that you find the most effective key to philosophical concepts. So, I would invite you to utilise poets on the subject of time in order to get what I precisely want to convey here: time considered not as measure but as duration is a conflicting notion in which duration is asserted indeed but also is denied. Aristotle has his own way of putting it. He says that time is more destructive than constructive.[1] Indeed, what distinguishes time as a way of duration is precisely the loss, the disappearance into the past, the destruction. In T.S. Eliot's "Four Quartets" there is a great deal about the philosophy of time. There is in it even a notion of timeless activity, of an activity which is not a motion. I do not know if Eliot found it in his reading of Aristotle—it is possible, he has read a great deal—or if it is just an effect of poetic intuition. Besides the "Four Quartets" there is quite a philosophy of time nicely expressed in innumerable works of poetry; I just have to mention it for all of you to realize that time as experienced in human existence is a commonplace of lyrical poetry of all times. I would say that the most useful poets philosophically are those who are less inhibited and who are not afraid to express sharp contrasts in blunt antitheses.[2]

In conclusion, I would urge you to be always careful not to confuse real time with a simulacrum of it, as did Kant. It seems clear that the time of which Kant speaks is the time of classical mechanics, of Newtonian physics. Now, except for ethical issues it is always safe to consider the philosophy of Kant as the exposition of the conditions which must be presupposed in order for a science of the Newtonian type to hold. I think that is definitely the problem of Kant. Thus, the reason he would not accept Humean scepticism is because it undermines Newton. But the kind of time presupposed by the equations of Newtonian physics is not real time; it is not the measure of real duration by the human intellect but a being of reason which, so far as I can see, no longer contains the before and after of real physical duration. It is rather like one more dimension of space. Thus, it should raise no obstacle to locate things at any point within it, and development in time should imply no change in time itself. The connection here with the principle of inertia is obvious. Since movement is a state just as well as rest, when a physical point is taken from a certain place to another place, it can also be taken back to the original place; likewise, movement from one place to another is interpreted in terms of time, but of a time which could never prevent the same physical point from getting back to the same place and the universe from coming back to the same state. Those are exactly the sort of postulates against which Meyerson protested as irrational, because they are obstacles to a rationalistic identification. Be that as it may, I think it is clear that the time of Kant is definitely a dimension patterned after the dimension of space. But that is not what real time is.

NOTES TO CHAPTER SEVEN

1. *Physics,* 221b 1.
2. Two interesting works on the general topic of time in literature are Hans Meyerhoff's *Time in Literature* (Berkeley: University of California Press, 1955), and Georges Poulet's *Studies in Human Time* (Baltimore: Johns Hopkins Press, 1955). See also *The Home Book of Quotations* (New York: Dodd, Mead and Co., 1934), for the *loci classici.*

Readings

J.F. Callahan, *Four Views of Time in Ancient Philosophy* (Cambridge, Mass.: Harvard University Press, 1948).

R.M. Gale (ed.), *The Philosophy of Time* (New York: Doubleday, 1967).

L.R. Heath, *The Concept of Time* (Chicago: University of Chicago Press, 1936).

D. Nys, *Cosmology* (Milawukee: Bruce, 1942), Vol. 2, Part 2.

Chapter VIII

PHILOSOPHERS AND FACTS

I

The Notion of Empirical Science

We contrast philosophy and empirical science. We contrast the empirical sciences with mathematics too. And the equivocal term "positive science" allows us to oppose both mathematics, admitted by all to be a rational [1] science, and the empirical sciences to philosophy.

Philosophy itself however appeals to experience; the greatest pride of a philosopher is to invoke in support of his conclusions the irrefutable testimony of facts. People talk of an empirical philosophy, and those who use this term to refer to their views do not intend to deny to philosophy its scientific character; this implies, then, that philosophy also is, in some sense, an empirical science. And yet, who would dispute that if philosophy is a science, it is a rational one?

Finally, mathematicians will on their own insist on the empirical origin of mathematical notions. From all this it follows that the terms "empirical science" and "rational science" can be extremely ambiguous.

Every science is rational in the sense which implies a certain preponderant use of reason. That science belongs to the class of rational knowledge and that it occupies within this class a very high position is so clear that we often use interchangeably the adjectives rational and scientific, taking them as equivalent

139

because of the manner they are opposed to the empirical. On the other hand, every science is empirical from a certain point of view. There is not a single idea in a man's mind that is not achieved ultimately through an abstraction from sensory data. A science that would owe nothing to experience, that would in no way be empirical, would be an angelic or divine science, not a human one. No matter what science we are talking of, at the very least we can say that it must be empirical from the point of view of the genesis of its notions.

The simple intellectual perception that is completed by the concept the mind utters as it exerts itself, leaves it in suspense; it is in the judgment, where the relation of truth is consummated, that the dynamism of thought reaches its terminal point. Now, the object to be grasped through the judgment is sometimes one of sensory experience, sometimes one of the imagination, and sometimes one that is intelligible only. Every science that studies the world terminates in judgments concerned with what falls under the senses; the object of a mathematical judgment can only be imagined; the object of a metaphysical judgment can only be thought. From the point of view of the judgment, which is the term of intellectual activity, every physical science, no matter what its type, is then empirical; the mathematical sciences are further removed from experience, and metaphysics is purely rational.

The mathematician is indifferent to facts, but the physicist and the metaphysician, to the extent that they can know them, can never be completely disinterested in them. Mathematical abstraction severs the relation to real existence that the object of thought had. Lines, triangles or circles can exist as pure mathematical essences only within the mind; as soon as we try to endow them with a real extra-mental existence we substitute for them something which is no longer mathematical but physical; in place of a straight line, for example, we put on paper with a pencil an area of graphite particles or we place on a blackboard a calcareous deposit with a piece of chalk. That is why mathematical explanations make use of neither efficient nor telic causes.[2] That is why there is no goodness in mathematical entities. And that is why mathematics is above all the realm of the necessary, which is the source of certitude: in mathematics everything is governed by the unconditional necessity of the order of essences; the necessity proper to the order of existence, with the conditionality which affects it whenever it is a question of created existences, has no place in mathematics. But as soon as an object of some science is capable of real existence, this fact is incorporable into the science and sometimes should be incorporated into it. For example, when the metaphysician treats of God, his labor would be in vain if he contented himself with an exposition of only the notion of the divine Being and of His attributes: he has to take a stand on an existential problem. Now, it is through sensory experience that we are aware of the existential status of the objects of our concepts; Anselm's and Descartes' theology is not conclusive because they did not understand this. "Through the senses it is established. . . ." said St. Thomas.[3] Thus, while the data of sensory experience are but a point of departure that is soon lost sight of and that cannot figure in the development of his science as far as the mathematician is concerned, facts are incorporable in the science of physics or of metaphysics. The enunciation of the fact of change plays a role in the demonstration of God's existence. Thus, from the point of view of the incorporation of facts into it, physics and metaphysics are empirical sciences while mathematics is not empirical at all.

Up to now we have used "physics" to mean any sort of systematic knowledge concerning the things of nature, or sensible being. If we raise the question of the relation of physical thought to experience from the point of view of the resolution of concepts, we shall be brought to distinguish two types of physical sciences, which cannot be called empirical to the same extent. Just as man's ontological structure has two poles, the spiritual and the material, so physics, which is of all the different modes of thinking the one most apposite for the conditions of human knowing, has two poles, the pole of being and the pole of the senses. Just as a man may through the way he orients his conduct ensure in himself the preponderance of the spirit or the preponderance of the flesh, likewise physics may, according to one's preferences, be organized in reference to the pole of being or to the pole of the senses. It will never be anything but a question of preponderance: the spiritual man is never free from his carnal body, and the carnal man testifies through the very depths of his debasement that he possesses within himself a power capable of the infinite; likewise, the physics which is the most systematically oriented to the maximum of intelligibility preserves within itself the notion of a sensible thing, and the physics which is the most systematically oriented to the sensible maintains a relation to being, which plays a formal role in every sort of thinking. Finally, just as when we project the course followed by the spiritual man we arrive at the realm of the pure spirits, whereas by projecting the course followed by the carnal man we arrive at the realm of pure animals, likewise when we project the course taken by the physics structured around the pole of being we naturally involve ourselves in metaphysics, whereas by projecting the course taken by the physics structured around the understood sensible we limit ourselves to the world of sensory data.

That is how we distinguish the two types of physics, the philosophical and the empiriological. Compare the definitions given for the same material object, for example, man, by philosophy and natural history, and note what you get when you resolve the terms of these definitions into clearer ones. The philosopher says: Man is a rational animal. What is an animal? A living thing that is sentient. What is a living thing? A being capable of giving to himself his proper perfection. All this is clearly metaphysical. The naturalist says: Man is a mammal that stands upright. What is a mammal? A vertebrate provided with special glands which secrete a liquid, milk, to nourish its young; milk is defined by its color, weight, and constituent chemicals; its constituent chemicals are defined in the end by observable properties, so that when the analysis is through, the ultimate illumination is provided by our sensations. Thus, from the point of view of the resolution of concepts we must distinguish within physics an empirical science which resolves its concepts into what is observable and a rational science, the philosophy of nature, which resolves its concepts into what is intelligible. Because of the different and contrary orientation of their concepts these two sciences resist fusion; despite the mind's tendency to unify all its knowledge it is impossible to form out of them a single organism as the philosophers of antiquity desired and as so many modern ones still dream of doing; on the contrary, it will even be forbidden to introduce into the structure of empiriological science any items borrowed from philosophy, and vice versa. The commingling of lights produces nothing but a shadowy spot. The empiriological sciences have maintained their progress by struggling against philosophy and the illusory explanations it provides as soon as it goes beyond its proper bounds; philosophy will be true to itself only if it resists the facile temptation to shift to empiriological science tasks which belong to it and to it alone.

The orientation of our concepts determines the kind of reasoning we shall use. The leading concepts of the philosophy of nature, being oriented towards what is intelligible, display necessary relationships which invite the use of deduction, as do also the concepts of mathematics and metaphysics. The leading concepts of empiriological science, being oriented towards what is sensible, can only imply necessary relationships which they do not display; that is why they do not lend themselves to be used deductively except under the guarantee of a previous induction, so that the preponderant role here belongs to inductive reasoning. Thus, from the point of view of the reasoning, the empiriological science of nature again merits the denomination of empirical science and the philosophy of nature, that of rational science. It is true that the science of nature unites itself, whenever it is possible, to mathematics to form a mixed science, one that is physico-mathematical or empiriometric; to the extent that it is clothed in a mathematical form the science of nature becomes deductive and, in virtue of that, rational. But we should note that this happens to it because of its mathematical form, not because it is a physical science.

To sum up, the philosophical sciences—here we are thinking

only of the speculative sciences of real beings—are empirical from the point of view of the genesis of their concepts, like every human science; from the point of view of their judgments, one of them, the philosophy of nature, is empirical and the other, metaphysics, is rational; from the point of view of their reasoning and of the resolution of their concepts both of them are rational sciences.

It is from the point of view of the relation of science to facts that we wish to examine the empirical character of philosophy. From this point of view there is no need to distinguish what is true of the philosophy of nature and what is true of metaphysics. Whether it is physical or metaphysical, philosophy is based on facts. But, what are these facts? Are the facts which philosophy uses to be identified with vulgar or with scientific facts, or do they have such specific characteristics that we have to define a philosophical fact as an irreducible species in the genus of facts?

II

THE NOTION OF FACT

"Everyone has a right to his opinions, but all reasonable men should be able to agree on the facts." Such is the usual view. Rarely will anyone be suspected of dishonesty because he refuses to grant to an opinion the importance which others attribute to it; we generally and with few reservations admit that in matters of opinion we are free to judge for ourselves and can decide in one way or another, no matter how paradoxically, and that this should not be considered a sign of malice.[4] But we admit no freedom for the mind in regard to facts; there is only a question of knowing if they were adequately established: in the presence of a well-established fact the mind can only give assent; the fact imposes itself on the mind. It is, in its own way, an absolute; as has been justly remarked, it is "the empirical absolute." [5]

That is why relativistic critics have had to persistently attack the notion of fact; having overthrown, at least in his own estimation, the absolutes of ontology and logic, the relativist could not consider his cause won as long as the absolute, any form whatsoever of the absolute, could find a refuge in experience, or as long as science and philosophy would admit along with the common man that any mind which is proud of its creative autonomy shall find its stumbling block in facts. The dilution of the notion of fact by showing that what we call a fact is but a product of the work of the mind—which by hypothesis is free and creative—that was truly, for a relativist, to utterly ruin the last citadel of the absolute. A good deal of zeal was used up in the project.

According to the relativistic criticism knowledge is a kind of constructive activity; what results from the work of the mind is always akin to an art object. Before attacking the absoluteness which is characteristic of facts the relativist denies that there is any essence existing independently of the mind that can impose its form on a concept. We shall not examine these postulates whose devastating gratuitousness has been so often pointed out; we shall hold firmly that the concept, to the degree that it represents a victory of the speculative intelligence, receives its characteristics from its object and not through the initiative of the subject. Thus, when we note in the formulation of what we call a fact the presence of a concept or a judgment, in a word of the results of a spiritual activity, this will involve on our part no incursion against the fundamental absolutism that the common man acknowledges to facts.

If a man has been killed on a street corner and the witnesses all agree, no doubt is permitted on that point; until the investigation establishes who the killer was and what his motives were, in all prudence we cannot yet speak of murder, for perhaps the one who struck the blow was just defending himself legitimately. But that there was a killing, that is a fact.

Suppose that a patient who had been declared tuberculous has died in a hospital and the autopsy confirms the diagnosis by discovering that the lung surfaces are greenish and pussy. The clinician notes, as a fact, the presence of caseous lesions.

The metaphysical mind, fascinated by its primary object, at

first wants only to know being and non-being. That is why Parmenides protested that it is impossible for being to be multiple and changing, that it cannot be but unique and fixed in the immutability of its eternal identity. But Heraclitus proclaimed the fact of change, and metaphysical thinking, no matter what the cost, will have to take this fact into consideration and be flexible enough to explain it.

We have here examples of a vulgar, a scientific and a philosophical fact. In each case there is an absolute that determines the mind's attitude; in each case it is an experience that resolves the issue. But this experience is not purely and simply an experience and this absolute is not simple either.

Suppose that when a magistrate interrogates them the witnesses declare they saw one man strike another and everyone admits that they are right. Nevertheless, strictly speaking, no eye has ever seen a man; it sees only a surface that is colored in a certain way, which is the object of the sensation in its purity. Around this pure sensation there organizes itself an extremely complex system of images, memories and intellectual elements that are so intimately involved with the sensation that we have no misgivings about stating, as a datum of experience, that we saw one man strike another. Let us try to distinguish the intellectual elements implied in the statement of such a fact and establish how they are related to the sensation.

Without doubt, when the man in the street declares that he has seen a man, he has in his mind a rather confused representation; but one thing that is sure is that there is something there. We would make him most uncomfortable if we asked him to state precisely what he means by the term "man," but there is no doubt that he intends it to refer to something which is a definite kind of being and which can be recognized through certain unmistakable characteristics. The man in the street has only a confused concept of "man"; while the botanist has a clear concept of *Fumaria officinalis;* but just as the botanist is not afraid of confusing what he means by "that which is not" with what he means when he talks of *Fumaria officinalis,* so the man in the street is equally unafraid of confusing what he means by "that which is not" with what he means when he utters the word "man." You cannot ask just anyone the difference between identity and otherness, but everyone knows what he means when he says, "this man and this other man." In the formulation of a vulgar fact then we use a concept; or better yet, two concepts are involved, to be united in a judgment: "A man struck another man," or "A man was striking another man." We do not have here a judgment in a question of essence, an affirmation of an identity whose extra-temporal character is expressed by a verb in the atemporal present, but a judgment of existence, in which the copula has a temporal character. Now, that which authorizes the one who proclaims a fact to affirm that there really exists in actuality a certain object of cognition or the union of two such objects, is the datum of sensation, around which the whole representational system enclosed in the formulation of the fact has organized itself. *To formulate a fact then is to make an existential judgment under the guarantee of a sensation.*

In every kind of judgment we shall find these three fundamental elements: an object of sensation, an object of intellection and an object of a judgment of existence. Now, what produces the diversity of types of facts is not the diversity of sensations. When the needle of a galvanometer moves, the layman and the scientist experience identical sensations, and yet the former knows only a vulgar fact and the latter knows a scientific fact. Nor is it the existential character of the factual judgment: at the sight of the same sensible phenomena both learned and ignorant persons think that *something* really exists. But for the one it is something vulgar, e.g., the movement of a mirror and of the light it reflects, produced by some unknown cause; for the other it is something scientific, e.g., the closing of an electrical circuit. In effect, within the structure of a fact the object of sensation plays only a material role; the formal side is taken care of by the object of the concept. But, all diversity of types results from a formal diversity. Thus, what distinguishes the different kinds of facts is the diversity of the concepts implied in the formulation of a fact; to a vulgar concept corresponds a vulgar fact; to a scientific concept, a scientific fact; and to a philosophic concept, a philosophical fact. The sensation calls forth the concept and the concept unites itself to the verb to be in an existential judgment whose expression constitutes the formulation of a fact. Whatever is the nature of the concept called forth, or whatever is the nature of the understanding's object that is discerned in the sensible datum, the fact that is formulated will be of the same sort. It all depends on the intellect's interpretation, what the intellect reads in the object of sensation. If the interpretation is of a vulgar type, there will be a vulgar fact; if of a scientific type, a scientific fact; if of a philosophical type, a philosophic fact.

We shall consider later on what is meant by a vulgar sort of thinking. We know however what distinguishes the philosophical type of thinking from the scientific (taking scientific in its modern and restricted sense in which we oppose the natural sciences, the empiriological and empiriometric sciences, to philosophy). Scientific thinking is oriented in the direction of the observable and measurable; philosophical thinking, in the direction of being *qua* being. In the presence of a factual utterance then, to tell whether we have to do with a philosophical or a scientific fact our task will come down to establishing the characteristics of the concept implied by the formulation of the fact, which characteristics are themselves understood by referring to the term we finally end up with in the resolution of the concept.

Pulmonary tuberculosis in an advanced stage of development is characterized by the presence of caseous lesions. We call a lesion caseous when its looks remind us of that of Roquefort cheese; this is surely the most clearly empiriological concept we could cite: what we use to define and name the caseous lesion, and to distinguish it from others, is first of all a sensible sign, of such a sort that if someone did not understand why the lesions present in a tuberculous lung deserve to be called caseous, the only way to make the term and the concept more clear to him would be to show him a slice of Roquefort cheese.

In the electrolysis of a saline solution the radicals move

towards the anode. "Radical" is the name we give to what is left over from a molecule of acid after we have removed the hydrogen atoms; "acid" is the name we give to a substance resulting from the combination of one or more positive hydrogen ions with a negative ion; an ion is an atom that has lost or gained some electrons; an electron is the smallest quantity of negative electricity so far known; electricity is a form of energy which manifests itself by such effects as the divergence of the gold leaves of an electroscope.

There is change in the world of our experience. What is change? It is the act of a being in potency precisely inasmuch as it is in potency. Act is being which is; potency is a non-being which is. We are in the presence of the first object of the metaphysical mind. Being is; non-being is not; there is nothing else.

Thus as soon as a fact is known the orientation of one's thought, whether it be scientific or philosophical, is clearly established; as soon as a fact is formulated the options for our thought are closed. It is then impossible that the fact which philosophy incorporates into itself be the same as the fact that science incorporates into itself; even if the datum of sensation was precisely the same the fact is not the same, because the concept involved in our intellectual knowledge of the fact is not the same. Consequently, *if it is true that it is impossible to integrate into the system of philosophical thought any item borrowed from the system of scientific thought, it will be impossible to incorporate into philosophy any scientific fact,* for a scientific fact connotes a scientific mind, with its characteristics which are opposed to those of the philosophical mind.

III

THE MYTH OF A PHILOSOPHY BASED ON SCIENTIFIC FACTS

Scientists excel in the defense of their independence from philosophers; usually a scientist holds them in scorn, especially those who want to speculate on natural objects. If, they say, philosophers devote themselves to the study of the spiritual or to the knowledge of themselves and of God, their games will perhaps be frivolous but will never hurt anyone. But in regards to natural objects, what could they say that was not arbitrary and figments of their imagination, they who are ignorant even of the fundamental facts without which a scientist would never urge anything? But science, being limited to its proper object, is not able to sate our curiosity concerning the world we experience, for what interests us above all is the being of things, the essential or causal basis of sensible reality; however, science, by considering sensible things as objects for observation and measurement, turns itself away from being. To provide our curiosity with the satisfaction it demands then, without giving ourselves over to the ruinous fantasies of the philosophers, we have only to erect science into a philosophy. This happens all the time, and should someone object that the scientist is not qualified to make pronouncements concerning philosophical problems, he can simply answer that his philosophy only expresses the final conclusions or hypotheses of positive science: it is a philosophy based on scientific facts. The notion has a magical prestige; nowadays most of the philosophical theories in vogue with the masses were spread by works synthesizing and popularizing science.

We ought not to be in too much of a hurry to proclaim the bankruptcy of scientism, for in spite of the renewal of interest in properly philosophical studies it remains true that the philosopher, in the eyes of the general public, presents a sorry figure compared with the scientist: he is a second rate citizen in the republic of scholars. Some philosophers have wanted to turn their humiliation ultimately to their profit. They have become science students and in order not to be accused any more of being arbitrary they intend to build up their philosophy on the latest scientific findings, as recorded in the proceedings of the most recent congresses. They will make, they also, a philosophy with a foundation of scientific facts and will thus merit applause twice, once as scientists and again as philosophers.

These confusions, which are sometimes ridiculous, manifest actually a point that is quite correct and extremely important, namely, the empirical concerns of contemporary philosophy. A "transcendental contempt" for facts is found less and less among philosophers, and one can only be glad of it. But if it is true that scientific facts cannot as such be assimilated into philosophy, if it is true that every scientific fact includes a piece of scientific theory which we cannot introduce within a philosophical system without diverting the course of philosophical thinking on a matter of the highest import, since it will be a point of departure, then the notion of a philosophy *based on scientific facts* would seem to provide only an illusory satisfaction of the exigencies of the concept of an empirical philosophy.

"At first sight," Bergson has written, "it may seem prudent to leave the consideration of facts to positive science, to let physics and chemistry busy themselves with matter, the biological and psychological sciences with life. The task of the philosopher is then clearly defined. He takes facts and laws from the scientists' hand; and whether he tries to go beyond them in order to reach their deeper causes, or whether he thinks it impossible to go further and even proves it by the analysis of scientific knowledge, in both cases he has for the facts and relations, handed over by science, the sort of respect that is due to a final verdict. To this knowledge he adds a critique of the faculty of knowing, and also, if he thinks proper, a metaphysic; but the *matter* of knowledge he regards as the affair of science and not of philosophy.

"But how does he fail to see that the real result of this so-called division of labor is to mix up everything and confuse everything? The metaphysic or the critique that the philosopher has reserved for himself he has to receive, ready-made, from positive science, it being already contained in the descriptions and analyses the whole care of which he left to the scientists. For not having wished to intervene, at the beginning, in questions of fact, he finds himself reduced, in questions of principle, to formulating purely and simply in more precise terms the unconscious and consequently inconsistent metaphysic and critique which the very attitude of science to reality marks out. Let us not be deceived by an apparent analogy between natural things and human things. Here we are not in the judiciary domain, where the description of fact and the judgment on the fact are two distinct things, distinct for the very simple reason that above the fact, and independent of it, there is a law promulgated by a legislator. Here the laws are internal to the facts and relative to the lines that have been followed in cutting the real into distinct facts. We cannot describe the outward appearance of the object without prejudging its inner nature and its organization. Form is no longer entirely isolable from matter, and he who has begun by reserving to philosophy questions of principle, and who has thereby tried to put philosophy above the sciences, as a 'court of cassation' is above the courts of assizes and of appeal, will gradually come to make no more of philosophy than a registration court, charged at most with wording more precisely the sentences that are brought to it, pronounced and irrevocable." [6]

This admirable page contains certain features peculiar to the Bergsonian system; we shall retain but the following thought, a common good of all philosophy: philosophy ought to entrust to nothing but itself the task of establishing the facts which it uses. By abstaining from any intervention in the establishing of facts, it condemns itself to submission to the rule of a mode of thinking foreign to itself, and this rule, not being suited to it, will devour its substance. Being the work of a scientist puffed up into a philosopher or of a philosopher transformed into a scientist, a philosophy *based on scientific facts* can consist of nothing but a sophistic philosophication of scientific views. It will be a rejuvenated scientism, a renovated scientism, nothing more. [7]

IV

COMMON, SCIENTIFIC AND PHILOSOPHICAL EXPERIENCE

Being exists; sensible being is multiple and subject to change; every sensible being presents a plurality of parts outside of each other; the things that fall under our senses admit of inequalities of perfection; there is order in the universe; all sensible beings are endowed with activity; there are some beings that are alive and others that are not; there are some beings that can know and some that cannot.

These facts, of inexhaustible fecundity for the philosophic mind, are not the results of any experimental technique; these fundamental philosophical facts are at the same time facts of common experience. We do not mean that they are equally evident for all men. Against anyone who would deny the existence of change we can bring up the testimony of the senses, but should anyone hold that there are not beings devoid of knowledge and that the things reputed to be such are to some degree capable of cognition and appetition, it would not suffice to invite him to open his eyes; sometimes, in opposition to a philosophical theory which denies a fact, the fact has to be philosophically established, but the philosopher's efforts then will not be directed to the conquest of new sensory data, inaccessible to the common man. The microscope, telescope and spectroscope would be of no help whatsoever in convincing a panpsychist of his error. Common experience furnishes us with all we can expect from the side of sensation; we have only to make precise the concepts implied in the enunciation of the fact and to show that the common data of the senses constrain us to unite these concepts in a given existential judgment. Thus, every philosophical fact from the data of common experience will not have the same dignity; those whose conceptual element represents an immediate intellectual reading of the sensible datum, previous to any technical elaboration, will enjoy a privileged condition. The first of these primary philosophical facts is the fact of the existence of being; there seems in second place to come the fact of plurality and the fact of change. An awareness of these facts imposes itself on the mind independently of all theories and it does not require the prior acceptance of any definition; the intellectual interpretation implied by their formulation results from a quite spontaneous mental functioning and if one disputes these facts, he will also have to dispute the efficacy of all mental functioning. Doubts in regard to the primary philosophical facts betoken the most advanced state of sceptical disintegration.

When we reflect on the reasons for the primacy of the philosophical facts which we call primary we understand why the philosophical facts that are based on the data of common experience themselves enjoy a priority in regards to the other facts that the philosopher will be able to use. In all rigor we can say that every essential part of the philosophical edifice is built on facts of common experience. This is because the fundamental theses of philosophy have highly general concerns and require the most perfect certitude, and it is common experience which presents us with the facts that are the most general and, by means of a correct critique, the most certain. There are no facts more general or more certain than the primary philosophical facts, and it is that that confers on them their character of primary philosophical facts and makes them have, in the order of experience, a dignity analogous to that of the first principles. In varying degree every philosophical fact answering to a common experience will share in this dignity, because of its certitude and generality.

We would not however hold that every philosophical fact is at the same time a fact of common experience. There are, as we shall see, some philosophical facts which can be established only through the technical elaboration of an experience. On the other hand, we should also note that the majority of vulgar facts are not philosophical facts. We can call vulgar any fact whose conceptual element does not involve any scientific elaboration, but there are many types of vulgar facts. Limiting ourselves to what is related to our thesis and without trying for a complete catalog, we can distinguish, in the affirmations of the vulgar mind, illusory facts, which any science will reject, facts of an empiriological sort and facts of a philosophical sort. It happens that an illusory fact is born of a precipitate intellectual interpretation. Thus, for the common man who is not at all aware of the results of scientific research, it is a fact that there is above our heads a solid, motionless vault. Another source of illusory facts: it often happens that the common man unites concepts that are themselves correct in an existential judgment that is founded on an insufficient number of observations. Animal stories are filled with such so-called facts, an example of which would be the suicide of scorpions. But there are some vulgar facts of the empiriological type which science incorporates into itself with hardly no changes in their formulation: it is a vulgar fact that alcohol kills and the biologist admits that the repeated ingestion of ethyl alcohol in excessive amounts brings about organic lesions and finally the death of the animal. There are likewise vulgar facts of the philosophical type which philosophy incorporates into itself. Common sense includes a rudimentary philosophy and this philosophy formulates facts which a technical philosophy will make its own while rendering their formulation more precise.

In their desire to counteract the idea of a philosophy based on scientific facts certain thinkers have implied that the matter of philosophical experience is entirely and exclusively provided by common experience. However, the very history of philosophy in its relation to the development of science testifies that certain scientific discoveries have affected, with the powerful decisiveness proper to the *empirical absolute,* the course of philosophical thought. The clearest example is that of the theory of celestial bodies. Struck by the impossibility of discerning in the appearance of the stars any qualitative variation, the ancients professed as a fact that the celestial bodies are incorruptible and that local motion is the only kind of change to which they are subject; being above all respectful of experience, Aristotelianism saw itself constrained to add to its general theory of the hylomorphic composition a rather disgraceful appendix, very little in harmony with the rest of the structure, to make a place in it for the illusory fact of the incorruptibility of the stars. The telescopes of the astronomers and spectrum analysis have forever rid philosophy of this romanesque phase. Shall we say that a scientific fact was incorporated into philosophy? Not at all, but a new philosophical fact was, whose subject matter, being inaccessible to common experience, was provided by scientific experience.

Light emanating from the sun, when analyzed with a spectroscope, produces a double black line that occupies the precise place of the double line of sodium; from this we conclude that there exists in the sun's chromosphere a substance optically characterized by the property of emitting when incandescent a light composed of two simple radiations with wave lengths of .5896 and .5890. That is the scientific fact. What philosophy will retain is that there exists in the sun a substance of the very same species as a certain sublunary substance and that consequently it is like it subject to qualitative and substantial change. Thus, the scientific form of a scientific fact has been stripped off it and its matter, a certain sensory datum, has been clothed with a conceptual form of a philosophical sort, thus making possible the formation of a philosophical fact, the only kind that can be assimilated into philosophy.

It appears that the contributions of science to the enrichment of philosophical experience are universally brought about under the conditions just described. Never does philosophy incorporate into itself a scientific fact, but it does happen that a point resulting from the pure sensory intuition involved in a scientific fact has philosophical bearings and can support a conceptual apparatus of a philosophical type. Take for example the progress made by philosophy, thanks to scientific investigations, in the theory of inanimate individual things. From the simple data of common experience the philosopher can declare with certainty that this animal here or that plant there (at least if it is a question of the higher forms of plant or animal life) constitutes one single individual, but to determine what is an inanimate individual, here common experience does not furnish us with adequate indications. Is this rock a single individual or an aggregate of individuals? On the basis simply of the testimony of common experience the philosopher can no doubt give to this question only a rather weakly probative answer. If limited to the macroscopic data of common experience, the philosopher could resist only with difficulty the temptation of thinking that organization, that is, the combination of heterogeneous parts distributed according to a certain type of order, is a privilege of living beings. For the ancients an inanimate individual was a homogeneous whole whose spatially distinct parts are of the same species and do not differ except by being outside one another. Finally, if we kept only to this same macroscopic experience we could only with difficulty abstain from affirming, as did the ancients, that the parts of any individual had to be in continuity, or at least in contiguity.

Contemporary research however forces us to hold, as a highly probable fact, that there exist molecules able, at least under certain conditions, of subsisting and acting in an isolated state. Science explains these molecules as complex structures of atoms that are specifically similar in the case of elemental bodies and specifically diverse in the case of mixtures; each atom is itself a complex whole composed of a motionless central nucleus, the proton, around which gravitate one or more ions. If enlarged a hundred thousand billion times, the hydrogen atom would appear to us as a ball with a radius of two decimeters describing a circle ten kilometers in diameter around an imperceptible little body having a radius of only a tenth of a millimeter. The definitions that science gives of the molecule, the atom, the ion and the proton lack any philosophical character; consequently the fact of the reality of molecules will not be able to be retained by the philosopher in its scientific form. What the philosopher will retain is the matter of the fact, which is to be found in the direction of the sensory intuition. No one doubts that no eye has ever seen a molecule and its different parts, but to the degree to which the scientific explanation of the experience is conclusive, it provides us with *the equivalent of a sensation*. It is likely that if our means of observation were sufficiently strengthened we would experience sensations that would suggest to the scientist the intellectual reading expressed by the molecular theory and to the philosopher the notion of an individual composed of parts that are heterogeneous and spatially distant. It is then a probable fact that an inanimate individual, at least in certain cases, is of a size imperceptible to the senses, that it constitutes a unified whole and that its parts are not contiguous—and this fact is a philosophical one obtained by a philosophical interpretation of sensations that are correctly suggested by scientific investigation, though not actually experienced. Although it is only a probable one, this philosophical fact will permit an appreciable progress of the theory of individuality in the philosophy of nature.

In this view, the independence of philosophy vis-a-vis science is fully safeguarded. Philosophy will not have to fear being reduced to the role of a registration court, for it will have intervened from the beginning in the determination of the fact and by this initial intervention it has taken out a positive insurance against being subjected to a foreign rule. But this view also maintains the possibility of a simply unlimited contribution from scientific experience to the progress of philosophy. For as soon as the refinements of scientific experience have once or twice made it possible for the philosopher to take possession of a new philosophical fact, there is no reason to doubt that a similar success will not be repeated a thousand times. Thus the empirical concerns of philosophers will be able to find nutriment that is continually replaced without there being any risk that the course of philosophical thought will be compromised. Thus, to the degree that philosophy will add to the old philosophical facts whose subject matter was provided by common experience new philosophical facts whose subject matter will be provided by scientific experience, a new empirical philosophy in no way kindred to any form of scientism will be able to constitute itself.[8]

In conclusion, we would like to point out what the general theory of facts suggests to us concerning the establishment of moral facts. Remember the principle we have been constantly invoking: the kind of fact we have depends on the kind of concept involved in the formulation of the fact. Whatever features the concept has, the fact will have the same ones. Now, it is enough to take a look at the features proper to moral concepts to conclude that being aware of a moral fact and formulating it will require conditions that are irreducible to the common conditions for being aware of physical facts. We might seem to be suggesting a tautology when we say that a moral concept is one having for its object a moral entity, but this apparent tautology is full of significance for whoever will take measure of all that is implied by the notion of moral entity. A moral entity is a kind of being that is absolutely *sui generis*, being defined by an essential relation of suitableness or unsuitableness with the ends of a free agent, so much so that it is impossible to know *what* a moral entity is without knowing *of what value* it is to this free agent. A botanist does not distinguish useful plants from noxious herbs, because to cheer up or to vex amateur gardeners does not at all help him to know the specific properties of a plant. But the one who studies moral affairs cannot dispense himself from knowing what value an act has in relation to the ends of the free agent, because moral affairs are constituted essentially by a good or bad use of our freedom. What remains of our notions of justice, temperance, theft or adultery if we abstract from the values they imply? If we forget about the idea of theft meaning an action adversely affecting the free agent in regard to the ends proper to him as a free agent and nevertheless try to think that idea, we shall have before our mind nothing but a representation denuded of any sense and capable of combining objects even more opposed than heaven and earth. There is no moral concept without a judgment of moral value and the knowledge of the fact will not be correct unless the moral judgment is itself correct.

If we can speak of facts in general, a fact is above all a certain individual reality's position as existing. In order to understand moral facts then, the rectitude of one's moral judgment will have to be secured within the same order of individual realities; it will not be enough to know what is right and wrong in general, one must besides be able to discern the goodness or evil of a given single action. But it is *prudence* that secures a precise determination of moral values on the level of the concrete, and prudence presupposes a virtuous will.[9] Thus, it is only the virtuous man who is qualified to understand moral facts perfectly. However, with the exception of mathematics, an awareness of the facts is the first step in scientific research. That is why the development of the moralist—and the true sociologist is, he too, above all a moralist—includes a purification, an asceticism, that is not ordered only to the intellect.

NOTES TO CHAPTER EIGHT

1. Philosophers contrast empiricism and rationalism: the former is an approach that emphasizes reliance on sensory data, while the latter emphasizes the use of insight and reasoning. We may similarly oppose empirical and rational science.

2. See Aristotle, *Metaphysics,* 996a 29 and 1078a 31.

3. *Summa Theologica,* I, q. 2, art. 3. Each of the arguments set up to prove the existence of God starts with the declaration of a fact: "It is established by the senses . . . For we find in these sensory data . . . or we find in things . . . For we find in things . . . For we see that . . ."

4. We do not mean that in the common view the logical necessity of ideas does not constrain the mind any; we only want to point out that the common view pays closer attention to the constraints imposed on the mind by facts.

5. G. Rabeau, *Réalite et relativité,* p. 140.

6. Henri Bergson, *Creative Evolution,* trans. A. Mitchell (New York: Random House, 1944), pp. 212–214.

7. For a recent discussion of the whole notion of different kinds of fact, from the point of view of an existentialist, see John Wild, *Existence and the World of Freedom* (New York: Prentice-Hall, 1963), Chapters 3 and 4.

8. Here we have to add a qualification concerning the use of experience in metaphysics. Strictly speaking, there are no metaphysical facts, because we experience only physical being. Nevertheless we call metaphysical any fact manifesting the existential status of a reality in which an intellect honed to analogical thinking will be able to recognize the physical analogate of a metaphysical reality. The being in question is above all sensible and material being, but it is in sensible and material being that the intellect honed to analogical thinking discovers being as being.

We may further note that if the philosophy of nature gets from the sciences an extension of its empirical material, it seems on the contrary that the empirical material of metaphysics is provided entirely by common experience: the facts of interest to metaphysics can only be of an extreme generality, as a result of which it is hardly probable that a technically elaborated experience will succeed in bringing forth any new facts for it.

162   THE GREAT DIALOGUE OF NATURE AND SPACE

9. George Sorel somewhere raises the question why research conducted according to Le Play's method has not given better results. It is because, he says, in order to penetrate into the most intimate depths of men *one must one's self be virtuous.* He adds that the reports of police officials in the past have not been very instructive.

Readings

* J. Kockelmans, *The World of Science and Philosophy* (Milwaukee: Bruce, 1969).

J. Maritain, *The Degrees of Knowledge* (New York: Scribner, 1959), Chapters II and IV.

E. Nagel, *The Structure of Science* (New York: Harcourt, Brace and World, 1961).

* W.R. Thompson, *Science and Common Sense* (Albany: Magi Books, 1965).

J. Wellmuth, *The Nature and Origins of Scientism* (Milwaukee: Marquette University Press, 1944).

* A.N. Whitehead, *Adventures in Ideas* (New York: Macmillan, 1933).

CHAPTER IX

SCIENCE, SCIENTISM AND REALISM

Permit me here to lay out, in all simplicity, the status of some research of mine, along with its conjectures and uncertainties. It is in this spirit of freedom, as far removed as possible from any dogmatism, that I shall take up two main aspects of the question under discussion: the problem of the unity of our knowledge of nature and the problem of the reality of scientific objects.

I

EPISTEMOLOGICAL PLURALISM

It is banal to point out that modern epistemology, to the extent especially that it remains faithful to the Cartesian ideal, opposes a monistic conception of knowledge to the pluralistic ideal that was characteristic of Aristotelianism. What is perhaps less noticed is that a more or less acknowledged epistemological monism often subsists within a pluralistic gnoseology. (I am using the term epistemology to mean the theory of science; by gnoseology I mean the more abstract and the more general theory of knowledge.) If we take scientism as it is usually typified, the extreme scientism whose strict representatives have without doubt been rare—at least among minds of any breadth—we must say that here epistemological monism becomes identical with an absolute gnoseological monism: all scientific knowledge proceeds or tends to proceed from a univocal intellectual light which is that of the positive reason. Here we see the first phase of the scientistic approach. It matters little, from the point of view we are taking, whether concretely we conceive of the positive reason in a more empirical or a more mathematical manner; what is essential is the univocity of the notion rather than its content. But scientism also claims to reabsorb into a univocal science all certain knowledge: apart from such a science there could only be conjectures, precariousness and fantasy. This means we have to give up religion and metaphysics and let positive science take over the domains of psychology, sociology, ethics, economics and political science.

If now we consider the reactions which have arisen in the last two generations against scientism, we note that most of them, with a considerable diversity of terminology, have in the end limited themselves to affirming the validity of certain kinds of knowledge that cannot be subsumed under positive science. The validity of religious knowledge, of metaphysical knowledge, of moral knowledge have been ardently proclaimed. But to none of this extra- or supra-positive knowledge is attributed a scientific character; this remains reserved for what we call positive science and thus scientistic monism is denied the second part of its claim but is granted the first. The development of Bergson's ideas seem to us to bear a rather remarkable testimony to the persistence of bias in favor of the univocal in modern epistemology. One could say that the whole gnoseological aspect of Bergsonism can be epitomized as an admirable effort to escape univocity, but Bergson's intuition escaped the framework of scientistic univocity only by resolutely placing itself outside the perspectives of demonstrative knowledge. Even though Bergsonian gnoseology may be pluralistic or at least dualistic, Bergsonian epistemology remains monistic. Consider also the attitude of the Vienna Circle: it manifests itself as far removed from the imperialistic spirit that characterized the heyday of scientism. "The groups represented here," declared Philip Franck at their congress in Prague in 1934, "are the last to overestimate the importance of science for life."

This is fine, but the speaker goes on to say: "We know perfectly well that human development is determined more by instinctive tendencies than by openly scientific ideas." Thus, unitary science—this is the very term which these Viennese use—is concerned with only a limited field in our universe, but what is outside of this field is given over to the domination of the instincts. Another member of the school, Rudolf Carnap, provides a clear interpretation of the incontestable fact that metaphysicians exist. The analysis of language has shown that metaphysicians exist. The analysis of language has shown that metaphysical terms are bereft of any meaning; so then, how to explain that in every age there have been men who devoted themselves to metaphysics? It is only because this illusory discipline provides an esthetic satisfaction that makes it a substitute for music. "Metaphysicians," Carnap concludes, "are musicians without any musical talent." [1]

In our view, whoever wants to work out a theory of the relations between philosophy and the sciences should above all take note of the scientific character of philosophy and understand that metaphysics, which is the archetype of all philosophical thinking, is at the same time purely and simply the archetype of all scientific thinking. If we often have to oppose science—in the restricted and modern sense of the word—and philosophy, we ought to do it with full awareness of what it involves and without ever losing sight of the fact that the most truly scientific sciences are of a philosophic sort. There are without doubt admirable analogies between philosophy, art and religion, but far from testifying against the scientific character of philosophy, these analogies testify in its favor. In fact, if the philosopher experiences some affinity for the artist or the religious man, it is because like them he spends his life in intercourse with mystery. Socrates' dictum remains always true: the great superiority of the learned over the ignorant man is that the learned one knows that he does not know, and the more science realizes its ideal by going deeper into its object, the more it becomes aware of its inadequacy and the more it acquires a feeling for mystery, through a marvelling which resembles that of the artist and foreshadows in a dim way the obscure face to face vision of mystical experience.

We have observed that in scientistic epistemology, be it of an absolute or limited sort, the exclusion of philosophy properly so-called or at least the negation of its scientific value are tied in with a univocal conception of scientific knowledge. If knowledge, wherever it is distinct from existence, that is, everywhere except in God, is a reiteration of being, then perfect knowledge, science, will be a perfect reiteration of being and therefore it ought to admit of as many different types as being itself does. Grant the realist conception of knowledge as the reiteration of being and grant further the fundamental doctrine of the analogy of being, you are at the same time granting the principle of epistemological pluralism. People have not paid enough attention to the import of the pluralist principle in the Thomistic conception of science, and too many believe that all has been said when one has pointed out that for Aristotle and St. Thomas the speculative sciences involve degrees of abstraction that are irreducible. If I may be permitted to introduce here a theological allusion, I would remind you that according to St. Thomas the most perfect human knowledge that we can conceive of in the state of our present life, the knowledge of Christ, is infinitely diversified: the human intellect of the Christ has as many scientific lights as there are essences to know. At just about the time when Descartes was writing his celebrated page of the *Rules* on the unity of our natural light, John of St. Thomas was again pointing out, at the beginning of his treatise on the division of the sciences, this great ideal of a multitude of sciences coinciding exactly with the multitude of scientific objects.

We mentioned a moment ago that modern epistemology, or, if one prefers, modern science in its theoretical exposition, *in actu signato,* has in general remained faithful to the monistic ideal of Cartesianism. Should we on the other hand consider modern science as lived, *in actu exercito,* it would seem that the evolution of science in the last three centuries presents on the whole the picture of a process of differentiation that continues without letup. It is as though the movement of history was dragging scientific thinking more or less successfully towards the pluralist ideal realized in the human intellect of the Christ. Thomists ought to be the first to rejoice over this event.

If we compare today's differentiated knowledge with the system of the sciences such as St. Thomas conceived of it, what above all strikes one's attention is the dissociation of the science of nature from the philosophy of nature. It is above all on the first level of abstraction, on the level of physical abstraction, that the process of differentiation makes itself felt. The distinction between the science of nature and the philosophy of nature seems to us to be a definitive acquisition of scientific and epistemological thought. Let us however hasten to note that this distinction does not itself preclude in any way ulterior differentiations. To affirm the distinction between the science of nature and the philosophy of nature is in no way to prejudge the question of the specific unity of the two sorts of sciences thus designated.

It is most instructive to try to uncover the reasons for all the resistance which this fundamental division of physical knowledge arouses from numerous Thomists. Up until recently, modern Thomists had the habit of incorporating the philosophy of nature into metaphysics, under the names of cosmology and psychology. It is not necessary to point out that from the point of view of St. Thomas' doctrine this assimilation of physical and metaphysical philosophy cannot be maintained for a moment. But is it merely a question of being unfaithful to St. Thomas? To our mind, it is a question of a confusion of viewpoints that is extremely prejudicial to the philosophy of nature, destructive to metaphysics, and well suited to make definitively obscure the problem of the relations between philosophy and the sciences.

Other Thomists, no matter what they think of the distinction between the philosophy of nature and metaphysics, find it repugnant to admit the validity of a nonphilosophical science of nature. One of them told me once that in his view modern science is a false philosophy, so that it would be better, if we had the time, to substitute for it a true philosophy; in concepts of the empiriological type, to go back to Maritain's terminology, this author saw only pseudo-concepts, bundles of images. This extremely interesting attitude which we could denominate an ontological integralism, presents some striking analogies with political clericalism. It is clear that it is above all in ontological notions that the intelligibility of being and the magnificence of truth manifest themselves, but does it follow that intelligibility is lacking in every nonontological apprehension of things? Likewise, it is evident that it is in a spiritual community that Christian life obtains the fullness of its bloom; does it follow that no temporal state can be Christian? Ontological integralism will vacillate between two attitudes. Sometimes it will attempt to violently take over areas which do not belong to it, and then we have philosophical pseudo-explanations of issues quite unsuited to philosophical explanation; we find a few examples of such pseudo-explanations in Aristotle and more numerous examples of it in the decadent scholastics, in Hegel and in the romantic *Naturphilosophie;* such conduct is analogous to clericalism's, as it seeks to absorb the temporal into the spiritual. Sometimes ontological integralism encloses itself in a splendid isolation and abandons to the imagination and to interior and utilitarian mental functions the universe whose conquest it has renounced, a bit like certain contemporary theoreticians who abandon to the devil terrestrial affairs and condemn as blasphemous the notion of a secular Christianity. In opposition to ontological integralism we have to make clear that despite the obscurity of empiriological notions they can provide a high degree of certainty. When I say "dog" or "cat," I am thereby expressing a true concept, even though it is obscure and cannot be made precise except by evoking sensations. In those passages of his writings in which he shows that a concept is fundamentally different from an image because of its essential relationship to being, Garrigou-Lagrange declares that he will borrow his examples, not from the level of the inductive sciences, but from the level of the sciences which attain their objects in their essences, no matter how progressively and incompletely they may do so, viz., mathematics and philosophy. Pedagogically this procedure is well justified: one should always start with the clearest cases. It would be proper today to carry out the complementary task of showing that the ideas that share the least in any metaphysical character and that are tied in the most with the data of the senses and the imagination, remain, they too, essentially relative to being. Although the bond that connects thought to being is more tenuous here than elsewhere, it is not broken and so long as any authentic science exists, it remains possible.

The division of physical knowledge into science and the philosophy of nature will, obviously, be rejected by all those who proclaim the vanity of ontological thinking. Because it seems to me to be the exact counterpart of the error we have qualified as ontological integralism, I shall cite especially the view put forth by Rudolf Carnap in his pamphlet *The Elimination of Metaphysics through Logical Analysis of Language*.[2] The terms which we qualify as empiriological, for instance, "arthropod," Carnap declares to be meaningful, because they are reducible to observations; the terms we describe as ontological Carnap declares devoid of meaning because no observation could ever uncover the object they signify. What is lacking here is the intuition of being and this is extremely serious. But if we brought to light the persistence of a bond between the intelligence and being in empiriological thought, we would at the same time be showing that the validity of empiriological thinking (which turns away from being even while allowing itself to be sustained by it) absolutely requires the validity of metaphysical thought (which devotes itself with all its strength to being), so that if the notion of being has no meaning, the notion of arthropod will not have any meaning either and as a result all science, along with metaphysics, will theoretically be reduced into a substitute for music.

The term "epistemological pluralism," which we have used, is justified by the analogical character of being, the object of all scientifique knowledge. Wherever we find a unity that is only analogical, predominance reverts to the pluralistic aspect. But since analogical relations preserve equally well a relative unity in the diversity of being, epistemological pluralism could never be an absolute pluralism. Through it are excluded all systems of strict compartmentalization or of multiple truth. Just as we cannot establish a bulkhead between philosophy and faith, neither can we place one between science and philosophy. And should it happen that materially identical propositions be true from the point of view of the scientist and false from the point of view of the philosopher or vice versa, it will still be necessary that the diversity of points of view be justified in the unity of a superior view which can only be philosophical.

II

THE REALITY OF SCIENTIFIC OBJECTS

The need to maintain the relative unity of knowledge along with its essential differences makes itself felt in a particularly pressing manner when we consider the problem of the reality of scientific objects. We were talking a moment ago about strict compartmentalization. The time is past, (thank God!) when Christian philosophers systematically abstained from verifying the conformity of their teachings to the dogmas of their faith. But for many people the bulkhead has only been displaced; instead of separating philosophy from faith they now separate science from philosophy. The progress is undeniable.

As soon as it loses its way, apologetics possesses a strange power of falsifying problems and introducing confusion into them. After a few apologists who were in too much of a hurry had heard some talk about the work of Poincaré and Duhem, we soon saw spreading among the faithful the view that modern science recognized itself to be incapable of grasping reality; people hoped in this way to be finished once and for all with all the problems involved with reconciling science with religious faith and its philosophical presuppositions.

It is useless to insist on the mischievousness of such over-simplications. But it would get us nowhere just to say that the question is a complex one: we would like to bring forth in orderly fashion some of its aspects.

By way of introduction we would note that one should not be in too much of a hurry to grant a literal philosophical sense to the declarations of scientists on the real import of their sciences. When a scientist utters such declarations—at least when it is a question of the whole of science and its principles—he is acting not so much as a scientist as a critic of scientific knowledge, that is, as a philosopher; from then on the weight of his judgment is measured not on the basis of his authority as a scientist but from the quality of his personal philosophy. If it happens that the scientist is doubling as an idealist philosopher, which occurs quite frequently, it will be proper to check whether his declarations on the real import of the sciences do not proceed, at least in part, from an idealist philosophy covered over with the pavilion of the sciences.

These precautions taken, it seems to me that to correctly place the problem of the reality of scientific objects we have to keep in mind the following principles of criticism:

1. Our mind can know the real but its grasp is never exhaustive. An object of knowledge is never anything but an aspect of the thing known, that lets remain behind it an inexhaustible depth of mystery.

2. The term "real" is analogical just as is "being." But, analogy involves not only a radical diversity but also inequality. The real is more or less real, just as being is more or less being. There is a real gradation in the reality of the objects of knowledge and we can affirm otherwise than metaphorically that the points of view adopted by the mind deliver to it unequally deep aspects of reality.

3. To affirm that the mind is capable of knowing the real and is made to grasp the real is in no way to affirm that every object of the mind really exists or can really exist. Besides the objects of the mind that are aspects of reality, there are beings of reason that neither exist nor can exist otherwise than in the mind.

4. Every scientific being of reason has a foundation in the real.

5. A fifth principle is concerned with the distinction between the scientific and the logical. We know how much this distinction has been clouded over through the whole of modern philosophy. For it would seem difficult to justify it from outside the perspective of critical realism. We shall say then that from within the perspective of this philosophy science is constituted by the second existence—in a perfect mode—of things in the mind. The things which exist a first time in nature—be it nature as actual or possible—exist a second time in the mind, and when they reiterate themselves there under conditions of essential perfection they there constitute science. But because of this second existence which they bear in the mind, things find themselves possessed of properties which neither exist nor can exist in nature, and it is these properties which logic takes as its object; in a word, it is they which constitute logic.

These principles being posited, we shall consider first of all the case of the ontological sciences, metaphysics and the philosophy of nature. It is clear that the objects grasped by these sciences admit of highly unequal degrees of reality. Nevertheless, no matter what the degree of reality of its object, theoretical philosophy is obliged to always express the real *just as it is,* even if it is in the least exhaustive way. This means that if philosophy makes use of fictions or of beings of reason, she will have to expressly declare her fictive constructs to be such and they will be of value only to the extent that they lead to the grasping of some real being. This is why the criterion of success, even though it may play some role in scientific matters, could never play one in philosophy; or rather, we should never consider any philosophical synthesis a success except that one which shall express the real exactly.

The grand error to avoid now would be to think that there is no reality outside of the ontological point of view. Such is basically the over-simplification which we believe we have recognized in the apologists of whom we were talking just now. Consider the privileged case of a reality which is susceptible of being defined at the same time by philosophy and by science, man, for example. The definitions are of unequal depth; that of the philosopher expresses a more deep and more real aspect of the reality under consideration; yet, the one and the other express something real.

The difficulties accumulate when the scientist is no longer considering the phenomenon in the proper sense of the word, the sensible forms and qualities, but rather the secret structures that are supposed to explain them. Even though theories of structure are nowadays great mathematical constructions, we believe that before mathematics influences the development of physical knowledge there arises a problem which ought to be considered independently of the special problems concerning the real import of physico-mathematical knowledge.

I shall content myself with calling your attention to the evolution of the concept of the atom in classical chemistry. At the beginning of the nineteenth century certain empirical laws led chemists to take back up for their own use, while giving it an entirely new meaning, the ancient notion of the corpuscular composition of corporeal things. New facts required incessant complications of the theory; such concepts as those of valences and of simple and multiple bonds made their appearance; molecular formulae were first developed on a plane, then in space; we succeeded in assigning a right and a left to stereochemical edifices, etc. Let us ask ourselves what there is that is real in all of that. No one contests anymore nowadays the real existence of molecules and atoms, but I doubt that many people consider stereochemical models to be the exact expression of a reality. (In saying "the exact expression" I mean an expression such that another expression that is incompatible with it would have to be held to be necessarily false.) Thus, at the origin of the theory there would be a physical reality; at a certain distance from this origin we find ourselves in the presence of an image of which we can neither affirm that it expresses a physical reality, nor deny it. It is either the exact expression of a reality or else it is a well founded fiction that *works*, that is, which saves the phenomena—"sozein ta phainomena."

I say a *fiction* and not a being of reason. For, a being of reason not only does not exist but cannot exist. If there is in the classical description of the atom a fictive aspect, is this fiction of such a nature as to be accompanied with an absolute impossibility of receiving existence? It would without doubt be rash, in many cases, to try to answer this question.

But here arises the problem of the epistemological value of well founded fictions, or of the sufficiency of the fiction. Police chronicles, or better yet detective stories, show us how many hypotheses can explain in an equally satisfying manner the appearances or the clues picked up in the investigation. Among the hypotheses which thus save the appearances there can nevertheless be but one which is true, and sometimes it is precisely the one that seemed the least plausible. A judge has an absolute obligation to identify the hypothesis which expresses exactly what happened and to exclude all the others. A fiction, no matter how well founded it is, possesses here no sufficiency whatsoever. And if the mind constructs a number of hypotheses which in the most favorable situation will all be false except one, the purely fictive hypotheses will have played but a transitory role; their whole value will have consisted in preparing one to recognize the true hypothesis.

Must we say the same about theories of structure? Or must we on the contrary say that the best one is whichever safeguards the appearances the best, no matter what the hidden reality is? Must we say that the fiction has the function of conducting us to the real and cannot substitute itself for it except provisionally? Or must we on the contrary say that the fiction suffices when it accounts for the appearances? Georges Urbain seems to affirm the "sufficiency of the fiction" when he declares that a scientific theory does not have to be true.

If it is permissible for us to hazard, with the most explicit reservations, a personal opinion, we would say that in the measure that the science of nature remains physical, it would appear to us to be compelled by its very nature to search out the real, to prefer the hypothesis that expresses the real exactly, to refuse all sufficiency to the fiction. The physicist *as such* would be comparable to the judge who ought to retain as the sole valid hypothesis the one that is the least plausible, the most onerous and accounts the least well for the clues, if some decisive proof such as the confession of the culprit establishes that things did come to pass in that way. When a writer like Georges Urbain seems to affirm the sufficiency of the fiction, it is permitted to ask one's self if fictions founded on the real have not become in his eyes a pure "quo," a pure means of attaining to the real.

I have said "the physicist *as such*." But the modern scientist is especially preoccupied with the mathematical interpretation of the sensible world. There is in him a mathematician whose particular exigencies do not involve physical reality. That is why the controversies concerned with the real import of science often have the appearance of a dialogue between the exigencies of the physicist's thinking and the "indifferences" of the mathematician's thinking.

I am here alluding to the doctrine of mathematical abstraction which the great commentators of St. Thomas have left us. While physical thinking and metaphysical thinking have as their object real being and use beings of reason only to know the real, and while logic considers the beings of reason whose laws are at the same time the rules of thought, mathematics, *precisely because of the abstractive process which confers its specificity on it,* finds itself placed in face of a universe in which a being of reason is equivalent to a real being. (Of course, it is a question here of a possible real, of what is able to exist outside of the mind, not of an actually existing real.) In the physical order and in the metaphysical order, the real enjoys in relation to the being of reason a double priority: of a priority that we can call causal, by extending analogically the notion of causality, in the sense that every being of reason is made in the image of a real being, and of a priority that we can call final, in the sense that the construction of a being of reason has as its function the knowledge of real being. In the mathematical order, the causal priority of real being remains, but its final priority is abolished.

We must insist on this point: it is precisely because of its noetic essence that mathematical thinking is subject to a law of indifference in regard to the reality of its object. It is not at all surprising then that it has a strange power of reducing to silence the realist exigency in every domain in which it will find it can apply itself. If we consider physicomathematical science in its structure as a mixed science, we have to say that from its formal side it tends to establish the equivalence of real beings and beings of reason. The "philosophy of as if" is its natural philosophy to the extent that it reduces itself to a pure mathematical interpretation of observable data. From this point of view, a concept that represents nothing else but a pure being of reason is for it preferable to a concept of the real that would express less well a possibility of measurement. Ultimately, if the mathematical form could impose an integral discipline on the physical content, we would, as in mathematics, be here in the presence of a system of absolute equivalence between real beings and beings of reason. But Meyerson's research has shown the resistive power that the physical content has.

Although our inquiry has been sketchy, we believe it suffices to show that the question of the real import of science can receive only a nuanced and not very glittering answer that is well suited to dissatisfy everyone, especially the popularizers. We would have to say that science attains to the real with certitude when it limits itself to stating empirical constants; in the theories set up to account for the empirical constants, everything leads us to think that there is there a solid nucleus of physical reality enveloped by a good deal of fiction; in the order of the fictive itself, we would have to distinguish at least *de jure* between fictions which have the character of beings of reason and those which do not; among the scientific beings of reason there would be a hierarchy to establish, according to how the being of reason character more or less profoundly affects the object which is thought—beings of reason are far from all of them presenting the same degree of unreality. All this could be summed up by saying that modern natural science undergoes at the same time the attraction of a properly physical ideal which requires that the real prevail over the fictive and that of a properly mathematical ideal which postulates the equivalence of the real and the fictive.

III

PHILOSOPHICAL EXPERIENCE

To conclude, we would like to attempt an application of the principles just sketched to the problem of the use of scientific facts in philosophy. According to an old and very enduring opinion, to establish the facts which the philosopher uses would be the function of the various sciences; the empirical basis of philosophy would be made up of scientific facts.

In one of the most beautiful passages of his *Creative Evolution,* Bergson has shown that if there does not exist an autonomous philosophical experience, there will not exist an autonomous philosophical thought either, so that a philosophy based on scientific facts will never be anything but a rejuvenated scientism.[3] In every enunciation of fact there are material aspects constituted by the data of pure sensation and a formal side constituted by the intellectual reading of the sensory data. Here as elsewhere, it is the form that determines the species or the type. That is why there is a typology of facts corresponding to the typology of concepts. That is why the same sensory input can be the occasion for the enunciation of a vulgar fact, of a scientific fact or of a philosophical fact: it all depends on the type of thinking involved in the perception and formulation of the fact. To incorporate the scientific fact *as such* into philosophy is then to introduce into the philosophical organism a strange body that is totally inassimilable and is therefore a source of perturbation, if it is true that all assimilation consists in the substitution of the form of the fed for the form of the food.

To want philosophy to be developed from a basis of scientific facts or to want it to be developed exclusively from the data of prescientific experience are basically the same error: in both cases there is a misunderstanding of the formal independence of philosophical thinking in interpreting the sensory universe. The truth is that philosophical thinking has to pursue the conquest of its empirical material, the hunt of philosophical facts, in every sphere of experience. Common experience is for it a privileged domain, but it is not the only domain open to it; certain scientific facts have a virtual philosophical import and it is the philosopher's function to disengage and actualize it, by giving a philosophical form to the empirical matter discovered within the formal perspectives of scientific thinking.

NOTES TO CHAPTER NINE

1. Rudolf Carnap, "The Elimination of Metaphysics," *Logical Positivism*, edited by A.J. Ayer (Glencoe, Illinois: Free Press, 1959), p. 80.
2. *Ibid.*, pp. 60–81.
3. This passage was quoted extensively in Section Three of Chapter Eight.

Readings

* Philip Frank, *Modern Science and its Philosophy* (New York: Collier, 1961).
* H.J. Koren (ed.), *Readings in the Philosophy of Science* (Westminster, Md.: Newman Press, 1958).
  W.O. Martin, *The Order and Integration of Knowledge* (Ann Arbor: University of Michigan Press, 1957).
* A.G. Van Melsen, *From Atomos to Atoms* (New York: Harber, 1960).

CHAPTER X

CHANCE AND DETERMINISM IN PHILOSOPHY AND SCIENCE

I

When a physicist speaks of chance he generally has in mind events characterized by unpredictability. If he belongs to the old deterministic school, he quickly remarks that no events are unpredictable in themselves and that the notion of chance would be altogether meaningless to an intellect capable of grasping all the causes at work in nature. Thus, reference to chance would signalize our failure to establish the complete set of factors which control an event. Chance would not enjoy any real existence. It would be but a name given to a certain kind of human ignorance. Such a view is not only found among physicists, it has also been held by many theologians.

Leaving aside the problem of the universal predictability of physical events, let it be remarked that the definition of chance in terms of unpredictability sounds congenial to common sense. Indeed, it is in the spontaneous usage of common sense that the physicists found it. We often use interchangeably such expressions as "by chance" or "by sheer luck" or "unexpectedly." We describe as fortuitous those accidents that nobody could predict, which implies that predictable accidents are not fortuitous and cannot be interpreted as chance events. The following is an example of a predictable accident: from a window I watch two cars moving in perpendicular directions; these cars are nearly at the same distance from an intersection and their speeds seem to be almost equal. It is early in the morning, traffic signals are not yet on, and neither of the cars is slowing down. At this time of the day a driver believes that he is alone on the roads. I anticipate the collision, first as threatening, then as inescapable. A small fraction of a second before the accident, my prediction is extremely well established. Plainly, no one would say that these uncautious drivers are victims of chance; rather, they are said to be the victims of their own carelessness, since they knew, or should have known, what risk they were running as they approached an intersection without slowing down and watching. On the contrary, if a driver on a mountain road was caught in a landslide, everyone would say that he was a victim of extraordinary bad luck.

In order to understand the philosophic significance of these interpretations, we must be aware that the expression "common sense" does not designate a function possessed of objective unity but rather an aggregate of notions which enjoy, on diverse grounds, the privilege of being accessible without any technical or scholastic training. The unity of these quite heterogeneous notions is not objective, but merely psychological. Thus, common sense contains an elementary philosophy which is, in a certain way, the origin of all philosophy and of all science; it contains a tyrannical imagery that science and philosophy often have to fight; it also includes a practical vision of the world which, though right and sound in its own field—i.e., in the practical field—may hamper the perception of theoretical truth. To test the common sense interpretation of chance, let us consider an example devoid of human significance and free from the disturbances that originate in practical concerns.

Imagine two players separated by a screen and unaware of each other's presence. An observer sees one of them roll a ball at a target; the same observer sees the other player roll another ball at another target. The balls collide. In a second phase, the screen is removed and the men are playing together. According to the rule of their game, they aim their balls so as to have them collide. Plainly, the first collision is a chance event, but it will not occur to anyone that in the second case the balls collided by chance. The second collision is not fortuitous because we assumed an agreement between the players. They have unified their action. The collision of the balls was present in their intentions. The first collision was fortuitous because the targets were diverse, the players unaware of each other's existence, and the actions nonunified. Predictability does not matter: the observer who twice saw balls rolling towards each other may have been able to predict encounter as safely in the first case as the second. The chance event is defined, not by unpredictability, but by the *nonunified plurality of the causal processes from which it results.*

This interpretation of the chance event as the result of an irreducible plurality of causes was expounded with remarkable lucidity by Augustin Cournot. In a celebrated page of the *Essay on the Foundation of Knowledge,* he points out that there exists in the universe of causes and effects, *inter*dependent and *in*dependent series. "It is not impossible that an event occurring in China or Japan may have some influence upon events happening in Paris or in London. But, in general, it is certain that the program a Parisian lays out for his day will not be influenced in the slightest degree by what is then going on in some city of China in which Europeans have never set foot. These are like two little worlds in each of which series of causes and effects can be observed developing simultaneously which are not connected and which exercise no appreciable influence on one another." [1] The distinction between interdependent and independent series makes it possible to define chance as follows: "Events brought about by the combination or conjunction of events belonging to series independent from each other are what we call *fortuitous events,* or the results of chance." [2]

We have suggested that the common sense interpretation of chance was governed by practical concerns. For, the predictability of a fact, even though it does not alter the relation of this fact to its real causes, deeply modifies its human meaning. A predictable fact, whether fortuitous or not, is a thing that human prudence must take into account. We do not say that the careless drivers asked for their misfortune: one of them wanted to go to work and the other wanted to go home. But no human responsibility is involved in accidents that are humanly unpredictable, whereas we may be held responsible for accidents that we should have been able to predict. Common sense defines chance in relation to prediction because it considers it from the standpoint of action. It is not surprising that philosophy has to correct common sense on a subject whose human significance never abates.

Yet, any discrepancy between the practical concept of a reality and its theoretical notion calls for explanation. If it is true that a predictable event which results from a nonunified plurality of causes remains a chance event in spite of its predictability, how is it that common sense attributes fortuitousness to unpredictable events alone? The answer is obvious: even though unpredictability does not pertain to the essence of chance, chance events, by reason of their complexity and irregularity, are generally harder to predict than events of nature. Inasmuch as a chance occurrence results from a nonunified plurality of causes, it cannot be predicted except on the basis of a complex system of initial data. On the other hand, in order to foresee that this bush will bear roses in the spring—except in the case of interference—all I need to know is that this bush is a rosebush. The irregularity of the chance event follows upon the nonunified plurality of the causes involved in its production: causes cannot operate jointly with regularity unless the regularity of their joint operation is guaranteed by the unity of a permanent principle.

II

Let us now try to see in what sense the notion of determinism and that of chance agree, and in what sense they exclude each other. It must be said, first of all, that the concept of an absolute chance, independent of all determinism, is a contradictory fiction. To see that contingency is grounded in necessity, it suffices to remark that the fortuitous encounter of two things in motion presupposes two things, each of which follows a nonfortuitous direction. The intersection of two lines presupposes two lines. A coincidence foreign to law and logic presupposes processes each of which obeys a law and has its own logic. Unless the philosophy of indeterminacy builds on an assertion of determinism, it is bound finally to deny the rationality of the universe and its reality.

The theory of chance not only presupposes the determination of the causal lines whose encounter constitutes the chance event, but also holds that the chance event, as soon as the causes from which it will result are actually posited, must occur inevitably, necessarily and determinately. But this determinism of the fortuitous is only *de facto* and not *de jure*. The necessity of the contingent is but an *inevitability,* a factual necessity; it is an *historical,* not an essential necessity. If you consider as factually existing, historically posited, an aggregate of causes whose actions converge without their convergence being embodied either in any of these causes or in the aggregate itself—which, by hypothesis, is nonunified—the result of their convergence, provided no free agent steps in, will occur necessarily (according to an historical necessity) without ceasing to be contingent and fortuitous: fortuitous, for it is not intended by any cause nor by any unified system of causes, and contingent, for it would not have occurred if another aggregate of causes had been factually posited.

This notion of a merely historical necessity, of a merely factual determinism, does not raise any difficulty so far as the demiurgical function of science is concerned. Engineers would not challenge the philosophy of contingency: they are interested in, say, regular telephone service between New York and San Francisco, and this is sufficiently guaranteed by historical necessity and factual determinism. But the theoretical ambition of science, its eagerness to explain, may bring about the denial of real contingencies and forcibly substitute a *de jure* determinism, an essential necessity, for merely historical necessities and merely factual determinations. For, historical necessity is not explanatory, and there is the rub. We now understand the meaning of the dialogue between Aristotelians and Stoics that St. Thomas recalls in his commentary on Aristotle's *On Interpretation*. Aristotle does not say, as does one misleading formula of the principle of causality, that all that happens has a cause and consequently is explained by a cause; he holds rather that the only being that has a cause is a *per se* being, i.e., a being provided with essential unity. The Stoics, on the contrary, assert that everything has a cause and although it may not be possible to explain an event through a single efficient principle, every event can be explained if a sufficient number of causes are brought to bear. These multiple causes, provided they are taken together, they held to contain the ground, the reason, the generating idea of the event to be explained. But this theory assumes a unified multiplicity and an explanatory causality. Plurality is not deemed irreducible: analysis, if carried far enough, will come upon causal unity.[3] Chance is eliminated and its proper mystery disappears. So, they conclude that all happens by necessity. The necessity spoken of is essential, and hence the event was written beforehand in the unity of a group of causes. This predetermination of the chance event is what Aristotle denies: ". . . not everything that happens has a [real] cause, but only *per se* being. What exists in merely accidental fashion does not have a cause, for, to speak properly, it is not a being. . . ."[4]

If we have well understood what is implied by the irreducible plurality of the causes whose interference brings about the chance event, we shall not hesitate to say that an event brought about by chance is an event without a real cause. It does not have a real cause because it has several real causes whose plurality is irreducible. A thing cannot be a real cause unless it has real unity. Now the plurality of the causes from which the chance event results has no unity except in our mind. We speak of it in the singular; we say *a* plurality, *an* aggregate, of causes, but it is the mind which supplies, here, a link that does not exist in the real world. The unity of this plurality or aggregate is a being of reason, just as much as a logical subject, a predicate, or an imaginary number. When chance is said to be a cause, let it be understood that the unity of this accidental cause is the work of the reason.

To say that the principle of causality rules out the reality of chance involves the assumption that if chance were real its unity as a cause would have to be real. A being of reason—the unity of chance as a cause—is imaginatively projected into the real world and, upon comprehending that it has no place there, chance is denied all reality. It follows from all this that a chance event is unexplainable, unintelligible and nonrational. This is why rationalism is intent on denying the reality of chance. Every explanation consists in an identification. To explain the Pythagorean relation is to show that, within a definite system of postulates, there is a strict identity between that which is called a right triangle and that which is defined as subject of the property under consideration. To explain an event by its efficient cause is to show that the features of the event are, at least in part, identical with the features of the nature described as its cause. When there is a question of a chance event, no causal identification is possible. Let us consider again the case of the uncautious drivers. One of them wanted to go to work, and the other wanted to go home. Suppose that no accident occurs. The first man actually reaches his place of work: this result is intelligible, explainable by reason of its identity with the representation which directed the driver and the car. The second driver actually gets home: this result is equally intelligible. But the collision does not resemble any causal antecedent, and it cannot be identified with the dynamism of any representation or nature. By reason of the very plurality of its principles, it is not identical with any of its principles. It involves an irreducible lack of intelligibility by reason of an irreducible lack of unity on the part of its causes. Hence, the rationalistic postulate of universal intelligibility, which requires the elimination of chance, requires also the elimination of plurality. Rationalism implies a theory of a universe that is absolutely one, fully lighted and fully actual. Diversity and change are reduced to mere appearances. But what becomes of the real universe if the reality of change and plurality is denied? Acosmism is the logical termination of the rationalistic endeavor. Thus, as Aristotle indicated, we would no longer be reasoning about things but about nothing.[5] Aristotle was referring to the monism of the Eleatics, but similarly Meyerson sees in the unitary and materialistic metaphysics of Parmenides the archetype of all attempts made at asserting the fundamental *identity* of things, be it at the expense of their *reality*.

Yet if often seems, in our daily life, that we succeed in explaining facts which, according to the preceding description, should be considered chance events. If this is an illusion, our illusion still has to be accounted for. Let us try to analyze one of those chance events that everyone thinks he can explain.

On a winter day, an old man crosses a street on a steep hill. He slips on the ice. An automobile going down the hill cannot stop in time: the man is killed. His death is unquestionably a chance event, for, as Cournot would say, there is no connection, no interdependence, between the series of the causes which determined the construction of the street on this hill, the series of the causes which led the old man to cross this street at this very time, the series of the causes which determined the formation of ice on this day and place, etc. However, people keep on reasoning. For the family of the victim, the accident is easily explained: its cause is the obstinacy of the old gentleman. He had been warned that the pavement was slippery and that he would be wise to give up his daily walk. But he did not listen. For the driver of the car, the accident is also easy to explain. It had occurred to him that he might be unable to control his car in such weather on such a steep hill, but he was in a hurry and did not detour. He considers with remorse that his carelessness caused the death of a man. For the head of the Street Department also, the accident is clearly explained: an employee was ordered to spread ashes on slippery roads but thought that he could afford to skip this side street with little traffic. . . .

But the meaning of these explanations is altogether practical. We would never have had the illusion that we could explain a chance event—by "explain" I mean to explain in the proper sense of the term, i.e., theoretically—if we had from the beginning possessed a clear notion of what a practical explanation is. Theoretical thought endeavors to know what things are, and in order to know them perfectly it needs to know them through their causes and in their causes, which is to explain them. Practical thought wants to know what we have to do in order to attain our ends, what we should have done in order not to fail of our ends, and what we must do in the future in order to stay along the line that leads to our ends. From the standpoint defined by the questions: what must we do? what should we have done? what shall we do?, to say that the accident is explained by the obstinacy of the old man, by the carelessness of the driver, or by the disobedience of the Street Department employee, makes sense. But these explanations are totally meaningless with regard to the theoretical question of the nature of things and of the way in which they proceed from each other. From this theoretical standpoint, it would be necessary to choose among the heterogeneous and unrelated causes that the family of the victim, the conscience of the driver, and the head of the Street Department assign to the accident. Does it lie in the obstinacy of the old man, the carelessness of the driver, or the disobedience of the employee? It is impossible to make a choice. Each of these causes is a necessary condition of the accident and, so far as theoretical reason is concerned, no one stands out. It cannot be said that the accident is explained by the whole sum of its causes, for these causes do not constitute a unified whole. Again, the illusions of common sense are traceable to the fact that an insight whose significance is entirely practical is mistakenly applied to a theoretical problem.

Moreover, the illusion that a chance event can be explained is soundly balanced by another feature of common sense psychology: the feeling of admiration and awe that the contemplation of a fortuitous event brings about in us. The bewilderment of the lovers thinking of the many encounters that were needed to bring about this masterpiece, their love, is an inexhaustible source of lyrical expression. Inasmuch as the causes of the fortuitous are several and are not unified, they explain nothing and leave us to our worries. Tired of seeking an explanation for the fortuitous event, the mind finds peace in the thought of Providence alone, and the bewilderment of chance, whether joyful or sad, often ends in an act of adoration. True, it is only on the level of the First Cause, on the level of the divine decree which organizes chance in a design whose ways are impenetrable, that the plurality of causes is finally unified. Should it be said that this final unification ultimately reduces the plurality of the causal lines, and that the acknowledgment of providential rule, as well as the assertion of universal necessity, amounts to denying the reality of chance? To this question it must be answered that the supreme distinction of divine government is precisely its ability to move creatures according to the mode of operation which befits them by reason of what they are. Under the influence of the First Cause, always infallibly effective, the natural event takes place naturally, the free event freely, the contingent event contingently, and the fortuitous event fortuitously. Concerning the real existence of chance, the problem is not whether the plurality of the causal lines is or is not finally unified, but whether this final unification takes place in nature or only in the Author of nature, who is transcendent to nature. The notion of natural necessity implies that the multiplicity of the causes involved in bringing about the fortuitous event is reduced *within* the physical world. The reality of chance disappears. But the theory of providential government holds that this ultimate reduction takes place only on the level of the First Cause, and thus the fortuitous remains, and retains all its mystery, though it is no longer burdened with dread.

III

Inasmuch as chance results from plurality, the field of chance grows as discrimination of plurality becomes keener. As we become able to perceive a greater number of factors, the part played by chance appears correspondingly greater. The microscopic approach to inorganic individuality was bound to multiply the conceivable interferences. The transition from the macrophysical to the microphysical standpoint had to be accompanied by a promotion of chance, since the mind was now confronted with a multiplicity unknown to the rough insights of common sense and of classical physics.

In a paper composed at the beginning of this century, Henri Poincaré wrote: "Perhaps it is the kinetic theory of gases which is about to undergo development and serve as model to the others. Then the facts which first appeared to us as simple thereafter would be merely resultants of a very great number of elementary facts which only the laws of chance would make cooperate for a common end. Physical law would then assume an entirely new aspect; it would no longer be solely a differential equation, it would take the character of a statistical law." [6]

Indeed, the transition from the "simple" to the "very great number" brings about the substitution of statistical laws for laws of causal description. But the integration of chance is what basically characterizes statistical law. For the first year student who repeats the experiment of Boyle, a gaseous mass has the character of an individual entity: one does not ask whether it resolves into a multiplicity of small masses each of which would act on its own as do the spectators of a play in case of fire. It is postulated that this gaseous mass acts as if it were one subject of existence and activity. Is chance thereby denied? By no means. The very formula of the law of Boyle refers to possible disturbances that the experimenter must prevent, e.g., temperature must remain the same, for a change in temperature would alter the result. But the so-called causal law, as it warns us against the substitution of a chance event for a process of proper causality, postulates that the elimination of disturbing factors is theoretically possible. However, if we consider that the undivided subject of existence and activity, i.e., the individual, is not the gaseous mass enclosed in this container, but a molecule so small that it would fit in any manageable container a billion times over, we no longer can study anything else than the general effect of billions of actions and interactions. Chance is now integrated in law.

Many thinkers have judged that this integration of chance in law constituted the most radical of all the revolutions ever undergone by the scientific mind and marked the definitive invalidation of the regulating ideal that science received from Greek rationalism.

At the dawn of Greek thought, the philosophy of universal mobility is represented by the school of Heraclitus and Cratylus. The universe of sense experience is a flux where natures lose their identity, for they do not contain any principle of stable existence. Delivered to the deceptions of a phenomenal universe where everything is in motion, early Greek thought goes through a phase of gloom. The phenomenon, inasmuch as it is carried by universal becoming, appears as the enemy of science. For Plato, a pupil of Cratylus, the scientific object must not be sought in the world of sense experience but in a supra-sensible universe of ideas and numbers. The phenomenal universe proves inconsistent as soon as it is subjected to theoretical analysis. It is merely an object of opinion. Whatever cognition we have of it is uncertain and contradictory: its real value is but practical. The scientific outlook implies first of all a conversion to the universe of things intelligible, and this conversion—here is the decisive mark of eternal Platonism—implies aversion to the flux of sense appearances. A science of nature is impossible.

Aristotle also starts off from the Heraclitean universe of perpetual change. But in this universal flux Aristotle perceives islands of stability, which are the sources of order. They are the universal natures of things, the types, the laws of how beings act and are acted upon and react. It is quite true that all things never stop changing, if by all things you mean every individual reality taken precisely as perceived by our senses. But within these changing individuals there is a universal set of characteristics which does not change. Looking at the sensible world the scientific mind ceases to follow an attitude of aversion in order to adopt an attitude of abstraction. The phenomenon becomes a friend of science; it leads the mind to the foundations of things. The phenomenal regularities, the constant relations between phenomena, in short, the apparent order within the observable world, reveal the intelligible order in which the mind satisfies, at least in the most favorable cases, its urge toward certitude and clarity. The leading idea of this philosophy of science can be expressed as follows: the chance event, inasmuch as it is not predetermined in any principle of activity, has the law against itself and consequently must have the character of an exception; the event of nature, inasmuch as it results from the undisturbed operation of a proper cause, has the law behind it and consequently must enjoy frequency. Existential and phenomenal regularities reveal the nature of things and lead the intellect to the perception of these eternal necessities of the possibles where the cause is identified with the thing and where the tendency to bring about a certain effect proves identical with a definite form of being. Thus, in the way of discovery, i.e., in the analytical or ascending way of science in the making, the apparent order—i.e., the observable regularity—leads to the intelligible order. Correspondingly in the descending and synthetic way, which is that of established science, the intelligible order—i.e., the necessary relations of abstract possibilities—accounts for the apparent order. The mind moves from order to order; it designates as the source of the order which is first for us, namely, the phenomenal regularities, the order which is first in itself, namely, the necessary constitution of the possibles.

Henri Poincaré refers to this concept of nature and of scientific explanation when he writes, in somewhat awkward terms: "It was either an immutable type fixed once for all, or an ideal that the world sought to approximate." But then he immediately remarks: "Newton has shown us that a law is only a necessary relation between the present state of the world and its immediately subsequent state. All the other laws since discovered are nothing else; they are in sum, differential equations." [7] Newton, then, would have been the first to work out this ideal form of the physical law.

Let us only remark that for Newtonian, as for Aristotelian science the apparent order of the universe, the phenomenal regularities, originate in an intelligible order. Minds trained in Newtonian disciplines were never tempted to place anything else than order at the root of order. This is established by the metapositive, nay, metaphysical character of determinism in the works of classical physicists. It is a certain vision of the intelligible order which commonly led the physicists of the classical period to deny chance, contingency, and freedom as incompatible with scientific determinism. The permanence of the Parmenidean myth in mechanistic systematizations expresses dedication to a basic pattern of intelligibility.

As a result of the microphysical approach, statistical formulas have been substituted for causal formulas in an increasingly large domain of scientific inquiries, but the temptation to place order at the root of order nevertheless continues to be felt by many minds. However, as Erwin Schrödinger remarks, Hume himself, in his celebrated discussion on causality, did not deny that some regularity obtains in the universe—but the very existence of such a regularity has been doubted in the last few years: "The basis of this skepticism is the altered viewpoint which we have been compelled to adopt. We have learned to look upon the overwhelming majority of physical and chemical processes as mass phenomena produced by an immensely large number of single individual entities which we call atoms and electrons and molecules. . . . The exact laws which we observe are 'statistical laws.' In each mass phenomenon these laws appear all the more clearly, the greater number of individuals that cooperate in the phenomenon. And the statistical laws are even more clearly manifested when the behavior of each individual entity is *not* strictly determined but conditioned only by chance. . . . If an initial state, which may be called a cause, entails a subsequent state, which may be called its effect, the latter, according to the teaching of molecular physics, is always the more haphazard or less orderly one. It is, moreover, precisely the state which can be anticipated with overwhelming probability provided it is admitted that the behavior of the single molecule is absolutely haphazard. And so we have the paradox that, from the point of view of the physicist, chance lies at the root of causality." [8]

To illustrate his position Schrödinger used the well-known example of life insurance: a company calculates with remarkable accuracy the percentage of death among its policy holders in a definite period, although it is impossible to foretell the future of any individual policy holder. Schrödinger says that two interpretations are possible. It may be held that statistics really expresses the essence of the laws of nature. If such is the case, the concept of causal connection must be rejected. But no experiment forbids anybody to prefer the opposite interpretation, according to which the behavior of each individual atom is determined by "rigid" causality, and statistical regularity is an ultimate result of the determination of elementary processes. The first position is "revolutionary" and implies that chance is primitive; the second is "conservative" and implies that chance is subjective. Schrödinger seems to think that it is up to philosophy to choose between the two.

The "conservative" position, Schrödinger says, will be preferred by those who were not convinced by the critique of Hume and hold that the principle of causality is a necessary law of thought. It will be held, with greater reason, by those who see in the principle of causality a necessary law of being and who consequently do not feel bound to attribute to chance a merely subjective signification.

IV

Before we consider indeterminacy in physics and its philosophic signification, it seems necessary to describe the diverse standpoints that a philosopher can adopt in a study of determinism.

The philosophy of determinism primarily pertains to the discipline called philosophy of nature or philosophical physics. The object of this discipline being nature itself, we ask, among other questions, whether the course of natural events is necessary or contingent, whether it involves both necessity and contingency, and—if it does—in what sense it is necessary and in what sense contingent? We attempt to answer these questions by the theories of change, causality and chance. In this philosophical interpretation of the physical world, all regulating concepts are suffused with an ontological meaning. Necessity and contingency and each kind of cause are defined in relation to being—more properly to changeable being—the proper object of the philosophy of nature. No wonder that the positive scientist feels puzzled by such expositions and needs an unusual amount of good will to follow them. He reads such words as "cause," "law" and "determination" and he does not recognize the meaning that they customarily convey to him. Sometimes he even runs into such frightening words as "being" and "essence" and feels strongly tempted to utter the adjective that the Vienna Circle has overused: *meaningless*. Let him however control his impatience and give credit for a while to philosophers. At least some of them hold resolutely that along with their ontological, or philosophical, science of nature, a nonontological and, in a way, anti-ontological physics is possible and necessary. These philosophers are convinced of the universal validity of such notions as necessity and causality. Yet they reject the silly theory that these notions could retain in a nonontological science, the meaning that they have in a science in which all is defined in relation to the being of things. When the scientist declares that he does not recognize, in the usage of the philosophers, the meaning that he is used to giving words, he merely illustrates the so often voiced truth, that the regulating concepts of all knowledge are primarily conceived—by common sense and by the philosophers—in relation to the being of things, and must undergo definite transformations before they can operate in a nonontological system.

The expression "philosophy of determinism" may also designate a chapter of the philosophy of science. Here the object considered directly is no longer nature in its necessity or contingency, but the concept of determinism and the ways in which we use it in order to express the real in our diverse conceptual systems.

When the physicist of today speaks of indeterminism, or indeterminacy, the first question for philosophic criticism is: How does he conceive the determinism in relation and opposition to which he speaks of indeterminism? It is up to the physicists to answer, and their answer, at first, seems to be unanimous: a process is determinate or, as they also say, causally conditioned, when and only when such and such ulterior phases can be predicted on the basis of a system of initial data. In a more technical language, the physicist speaks of a deterministic system whenever, by the very fact that we know the position and velocity of a material point at the initial instant t,[1] we are able to calculate its coordinates at the ulterior instant t. Thus, determinism is conceived as a possibility of certain and exact prediction.

If such is the definition of determinism, an absolute determination of scientific phenomena would require absolutely precise measurements as well as an absolute isolation of the factors under consideration. As Louis de Broglie writes, "In practice however, there invariably arises indeterminateness that is in a sense accidental, since it is due to the imperfections of our methods of measurement. Actually, the co-ordinates and the initial velocity of any moving body are never known with absolute precision; all that we can say is that they fall within certain limits, generally very narrow, which gives the exactness with which these magnitudes have been measured. From this slight indeterminateness of the initial data, however, there follows an indeterminateness in our previsions of the positions and final velocities of the moving body, this indeterminateness generally increasing with the lapse of time. But I repeat that according to classical notions this indeterminateness is contingent, so that it ought to be eliminated completely if only we succeeded in progressively perfecting our methods of measurement." [9]

Let us insist upon this point: this indetermination, because of its merely accidental character, does not raise any question such as would involve principles. Physics entered its "crisis of indeterminism" when a new kind of indetermination was evidenced, a kind which must be termed essential—in relation, of course, to the definition of determinism as the possibility of certain and exact prediction. Thus de Broglie continues: "Now the new concepts introduced by contemporary physicists are the following: Beginning with the idea that every observation necessarily introduces some degree of disturbance into the phenomenon under investigation, they conclude, on the basis of an acute analysis, that even if we possesed infinitely perfect measuring instruments, it would be impossible to ascertain simultaneously and with absolute exactness both the position and the velocity of a corpuscle; quite apart from the contingent indeterminateness already referred to, there would always be an essentially irremovable indeterminateness." [10]

In brief:

1. There is in microphysical observation a margin of indetermancy, and it is not accidental.

2. The cause of this indeterminacy does not lie in the relation between the initial and the ulterior state, but in our inability to obtain simultaneously all essential data concerning the initial state.

3. This impossibility follows upon the disturbance necessarily brought about by the observer.

4. This disturbance is rendered inevitable by the corpuscular nature of light. One must illuminate what one wants to observe, but to illuminate an electron is to throw at it a corpuscle which causes it to deviate from its trajectory—the Compton effect.

What are the philosophic consequences of this state of affairs? This question can be considered from the standpoint of the philosophy of nature or from the standpoint of the philosophy of science. From the first standpoint, it becomes: Does microphysics contribute any discovery of such significance as to force a reform upon our view of physical necessity? This question seems to call for a negative answer. Philosophy asserts that the event of nature, inasmuch as it is predetermined in its proper cause, enjoys a *de jure* necessity and can be foreseen in its cause. Such foreknowledge, however, is subject to two reservations: we may not be able to penetrate the cause so completely as to perceive in it the predetermination of its proper effect, and we can never remove the contingency of disturbance with absolute certainty. However, the chance event, which is not predetermined in any cause, which does not enjoy any *de jure* necessity, and which consequently cannot be the object of scientific inquiry—for science is cognition through proper causes— still can be predicted in an historical sense by reason of the *de facto* determinism to which it is subjected. Again, the possibility of predicting is qualified by two reservations: a chance event cannot be safely predicted unless the nonunified system of the initial data from which it results is exhaustively known; moreover, interference by a factor foreign to the system of the initial data can invalidate even a well-grounded prediction. Both in the case of the chance event and in that of the event in nature, the theoretical conditions of an absolutely certain prediction are never satisfied, and scientific prediction will always be affected by some uncertainty, which uncertainty can be indefinitely reduced by improved methods of investigation and a more complete isolation of the investigated systems.

The essential indeterminacy of which the new physics speaks seems to boil down to the fact that in some domains of natural inquiry observation necessarily disturbs the observed, so that the first condition of any rigorous prediction, namely, the accurate establishment of a system of initial data, is impossible. Thus we must give up applying to all the things of nature the *scientific* principle of causality and the scheme of scientific prediction. The recently revealed fact of indeterminacy concerns the relations between man and nature rather than the course of natural events. So, the important problem raised by the indeterministic ideas born of quantum mechanics does not pertain to the philosophy of nature: it rather pertains to the critique of scientific knowledge.

V

It is well-known that Max Planck, whose discoveries in the discontinuous nature of light are at the origin of the new physical conceptions, never ceased to oppose the indeterministic interpretations of some of his colleagues. Planck is not only a great physicist; his studies on the problem of determinism evidence a profound philosophic sense.

Let us review the main points contained in his remarkable article entitled "Causation and Free Will." [11] He holds:

1. There is no sharp separation between science and philosophy; there is a *philosophy within science*.

2. This philosophy is realistic and holds that a thing independent of the mind is the ultimate measure of the mind. Planck goes so far as to say that the ultimate goal of science is to know things *in themselves,* which seems to us equivocal and rash.

3. The principle of causality is independent of all sense observation.

It follows that the impossibility of expressing certain events in deterministic formulas does not weaken at all our belief in the universality of real determinations. This belief results not from any definite observation, but from a metaphysical requirement. On the nature of this requirement, Planck expresses himself rather confusedly; he may well attribute to it a certain degree of irrationality.

Under these conditions, what becomes of the definition of determinism as the possibility of certain and exact prediction, which seems to be unanimously accepted by the physicists of our time? Planck does not disrupt the unanimity, but he soon declares that this definition of determinism has but provisional significance. The observable and measurable determination which constitutes the possibility of certain and exact prediction is but the empirical effect and the sign of a deeper determination pertaining to the things themselves as independent of our minds. And if it is true that the first action of scientific thought is a jump into the metaphysical, it must be said, correspondingly, that this determination pertaining to the things themselves, this ontological determination, is the objective foundation of scientific certainty.

The meaning of the controversy between deterministic and indeterministic physicists is now quite clear. It does not seem to be exceedingly daring to assume that physicists are agreed on the facts materially considered. Their final divergences proceed from diversity in their epistemological conceptions, i.e., from different ways of conceiving the relation of science to the being of things. Some of them consider that a proposition cannot have a scientific meaning unless it refers to observations and measurements that can be really effected according to definite rules of operation. These want an entirely disontologized science, a science exclusively related to the observable and the measurable, without any conceivable reference to the being of things. For them, to say that the empirical or, better, empiriometrical determination is but the criterion of a real determination, is a proposition devoid of scientific meaning; they hold that a real determination which would be distinct from the empiriometrical determination does not exist for scientific thought. From a scientific point of view, a determination modified by such ontological reference is like a quantity multiplied by zero. Every scientifically meaningful determination is empiriometrical, and when empiriometrical determination is defective, we can talk only of indeterminacy.

By contrast, for Max Planck and those of his spirit, the assertion of determination retains its scientific significance in spite of the restrictions imposed, in the new ways of physical research, upon the possibility of predicting events with certainty and exactness. Science is not conceived by them as a system of entirely disontologized cognitions but rather as an empiriometry supported by an ontology. The problem of the philosophic meaning of physical indeterminism ultimately resolves into the question whether positive science ought not to free itself from every ontological implication and from every reference to being.

In the examination of this question, it would be fitting to use first the method illustrated by Meyerson in his study of the products of scientific thought, in which he seeks to disengage the rules spontaneously obeyed by scientists in the actual construction of positive knowledge. Much of the material for such an inquiry is found in the works of Meyerson. An investigation of it would bring forth two main facts. On the one hand, positive science pursues indefatigably the work of its own disontologization; it endeavors to be no longer a philosophy of nature. As we reread Newton, we are astonished to notice that definitely ontological statements loom large in this Newtonian science which was considered by generations as the accomplished type of positive knowledge. Our bewilderment gives an idea of what a long way it is from Newton to our contemporaries. On the other hand, we notice that the thinkers the most determined to purify science of all ontological residue run into apparently insuperable resistance, and we understand that their undertaking, if it could be carried to the term expressly aimed at, would annihilate science and annihilate itself at the very instant of its triumph: the *absolute* rejection of a reference to being has the disadvantage of suppressing the possibility of speaking and thinking.

One might have suspected that the case was such if one had paid better attention to the metaphysics of understanding. A second method, designed to supplement the preceding one, would start from the elementary analysis of intellectual operations and, through a process of successive differentiations, would lead us from the general critique of the understanding to the general critique of the particular ways of the scientific intellect in each particular science. Let it be said that the movement of positive science seems to be animated by antinomic tendencies, both of which are sound and whose conflict is an indispensable factor of progress. One of these tendencies aims at the ontological interpretation of nature and is common to science and to philosophy. The other pertains specifically to positive science: it aims at expressing every scientific object in terms of observations and measurements. The first tendency is constantly repressed by the second, and during the phases of smooth development it remains mostly unconscious. But as soon as a crisis occurs, physicists are caught philosophizing.

But if it is true that there exists a philosophy within positive science, we must conclude that the philosophy of nature admits of two states. It exists in a state of disengagement, of clarity and of consciousness in the discipline which bears this name of philosophy of nature and is the work of the philosopher. It exists in positive science obscurely and vitally. The philosophy of the philosopher receives from the philosophy of the scientist healthful stimulation. Conversely, the philosophy of the scientist cannot disengage itself from the state of obscurity in which it is kept by the pressure of positive requirements and become critically conscious of itself, unless it becomes a philosopher's philosophy and submits to the general laws of philosophic disciplines.

Finally, if we consider, from a psychological standpoint, that which takes place in the subjectivity of the scientist and in the scientifically minded public, attention should be called to the great part played by a factor not reducible either to the positive or to the ontological component of positive science. This third factor pertains to the imagination; it is nonrational, though in a constant relation of mutual influence with the rational processes which it accompanies and by which it is penetrated. We propose to call it one's *cosmic image*.[12]

Every great scientific epoch is characterized, for the historian of civilization, by the predominance of a certain cosmic image. Thus, in biology, the Linnaean science is characterized by the image of everlasting patterns, whereas the image which illustrates the Lamarckian and Darwinian epoch is that of patterns that change unnoticeably. The big error of many interpreters is to confuse the cosmic image by which science is accompanied in the subjectivity of the scientist with the philosophy immanent in science. Revolutions in the properly positive part of positive science always entail a change of cosmic image, but they do not necessarily entail corresponding revolutions in the philosophic foundations of science. Thus, the changes which recently took place in the positive systematization of physical knowledge as a result of the impossibility of applying to micro-phenomena the scheme of certain and exact prediction altered radically the cosmic image which haunts the minds of the physi-cists. Yet it does not seem that it should modify in essential fashion our ontological vision of nature, in which the concepts of substance, causality and finality are fundamental.

NOTES TO CHAPTER TEN

1. Augustin Cournot, *Essay on the Foundation of Knowledge,* translation and introduction by Merrill H. Moore (New York: Liberal Arts Press, 1956), p. 40.
2. *Ibid.,* p. 41.
3. Marcus Aurelius, *Meditations,* VII, 9.
4. Thomas Aquinas, *Commentary on Aristotle's On Interpretation (Peri Hermeneias),* trans. J.T. Oesterle (Milwaukee: Marquette University Press, 1962), I, lect. 14.
5. *Physics,* 185b 17–25.
6. Henri Poincaré, "The Future of Mathematical Physics," *The Foundation of Science* (New York: Science Press, 1921), p. 319.
7. *Ibid.,* p. 292.
8. Erwin Schrödinger, "The Law of Chance," *Science, Theory and Man* (New York: Dover, 1957), pp. 41–3.
9. Louis de Broglie, *Matter and Light; The New Physics,* trans. W.H. Johnston (New York: Dover, 1946), pp. 227–8.
10. *Ibid.,* p. 228.
11. Max Planck, "Causation and Free Will," *Where is Science Going?* (New York: Norton, 1932).
12. A.N. Whitehead graphically described what we refer to as cosmic image in *Science and the Modern World* (New York: Macmillan, 1928), p. 11.

Readings

* E.A. Burtt, *The Metaphysical Foundations of Modern Physical Science* (New York: Doubleday, 1952).
D. Bohm, *Causality and Chance in Modern Physics* (New York: Van Nostrand, 1957).
E. Harris, *Nature, Mind and Modern Science* (New York: Macmillan, 1954).
* J. Maritain, *Challenges and Renewals* (New York: Meridian Books, 1968).
* P.H. Van Laer, *Philosophico-Scientific Problems* (Pittsburg: Duquesne University Press, 1953).

OTHER TITLES

206